RESEARCH AND DEVELOPMENT

as a Determinant of U.S. International Competitiveness

BY RACHEL McCULLOCH

Harvard University

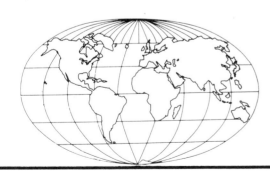

NPA Committee on
Changing International Realities

Research and Development as a
Determinant of U.S. International Competitiveness

CIR Report #5
NPA Report #161

Price $3.00

ISBN 0-89068-044-2
Library of Congress
Catalog Card Number 78-63432

Copyright October 1978
by the
NATIONAL PLANNING ASSOCIATION
A voluntary association incorporated under the laws of
the District of Columbia
1606 New Hampshire Avenue, N.W.
Washington, D.C. 20009

79-535

Contents

Research and Development as a Determinant of U.S. International Competitiveness

by Rachel McCulloch

NPA's Committee on Changing International Realities and Their Implications for U.S. Policy — inside front cover

A Statement by the Committee on Changing International Realities — vi

Members of the Committee Signing the Statement — ix

Chapter 1 — **Introduction** — 1

Chapter 2 — **The State of U.S. Technology** — 5

An Overview of Current U.S. Expenditures for R&D — 5
How Does the United States Compare with Other Countries? — 8

Chapter 3 — **R&D in Relation to Productivity Growth and Competitiveness** — 16

Effects of R&D on Productivity — 16
R&D as a Determinant of Trade — 20
General Equilibrium Effects of Expanded R&D — 24

Chapter 4 — **Government Policy toward R&D** — 27

The Need for Government Action to Promote R&D — 27
The Policy Spectrum — 29
R&D Programs of U.S. Trading Partners — 32
U.S. Technology Enhancement Programs — 34
International Comparisons — 35

Chapter 5 — **Conclusions** — 38

Bibliography	**43**
National Planning Association	**47**
NPA Officers and Board of Trustees	**48**
Recent NPA International Publications	inside back cover

TABLES

Chapter 1

1–1. Absolute Resources Devoted to R&D in OECD Countries ... 2

Chapter 2

2–1. U.S. Federal R&D, Fiscal Year 1976 ... 6
2–2. Performance of Research and Development, by Sector, 1976 ... 7
2–3. U.S. Industrial R&D, 1975 ... 7
2–4. R&D as a Percent of GNP, 1963 and 1973 ... 8
2–5. Scientists and Engineers Engaged in R&D, 1963–73 ... 9
2–6. Percent of Government R&D Allocated to National Defense, 1961–62 and 1971–72 ... 10
2–7. U.S. Patent Balance, 1966–73 ... 11
2–8. U.S. Net Payments from Patents, Manufacturing Rights and Licenses, 1960–74 ... 12
2–9. GDP per Employed Civilian, 1960–74 ... 13
2–10. U.S. Trade Balance for Selected Commodities, 1960–75 ... 13
2–11. U.S. Balance of Trade by Region, 1975 ... 14

Chapter 3

3–1. Rates of Return on R&D for Individual Innovations ... 17
3–2. Rates of Return on R&D for Individual Firms ... 18
3–3. Rates of Return on R&D for Manufacturing Industries ... 18
3–4. Hourly Compensation of Production Workers in Manufacturing, 1960–76 ... 21
3–5. Hourly Compensation of Production Workers by Selected Industry, 1976 ... 22

Figure 1. Direct and Indirect Consequences of Policies to Promote R&D ... 25

TABLES

Chapter 4   4–1.   Government R&D Programs of U.S.
                        Trading Partners   33
                4–2.   Percent Distribution of Government
                        R&D Support, by Function   36

Appendix   A–1.   Allocation of Funds for Basic Research,
                        by Source and Performer, 1976   40
                A–2.   Allocation of Funds for Applied Research,
                ∙   by Source and Performer, 1976   40
                A–3.   Allocation of Funds for Development,
                        by Source and Performer, 1976   41
                A–4.   R&D Expenditures as a Share of GNP,
                        by Country, 1963–73   41

A Statement by the Committee on Changing International Realities

Research and development (R&D) play a crucial role in modern societies. They are the basis for economic growth and hence for the continuing improvement of living standards, which is a major social goal. Moreover, they are an essential aspect of the export capability of the advanced industrialized nations and hence of their ability to import the goods and services required to help maintain and improve their living standards. This is especially true of the United States, whose export capability depends importantly on its leadership in high-technology products. Preservation of that leadership in turn requires continuing substantial research and development and the capital investment needed to apply their results.

During the 1950s and for much of the 1960s, the United States devoted a sizable percentage of its resources to research and development, which amounted to 3.0 percent of GNP in the mid-1960s. However, in the late 1960s, the proportion began to decline, from 2.9 percent in 1968 to 2.3 percent in 1976. This drop was initially due to the absolute fall in government-financed R&D after the peak of the space program. However, industry-funded research and development has declined in the 1970s.[1] And, while R&D was declining as a proportion of GNP in the United States, it was increasing in other countries, particularly in other major trading nations, such as Japan and Germany. By 1973, the average level of technology in Japanese—and probably also in German—industry exceeded the average level in American industry.[*,2,3] The implications of the U.S. lag have become all the more serious in the last few years as the deficit in the U.S. balance of payments has been mounting. Reducing the deficit to tolerable levels in the years ahead requires increasing U.S. exports, which depends in significant measure on the ability of the United States to maintain and improve its comparative advantages for high-technology products through adequate R&D. It also means that, wherever possible, the United States must increase the productivity of domestic industries whose

---

1   *I believe the statement that industry-funded R&D has declined in the 1970s is incorrect.* —**J.S. Parker**

*See Dale M. Jorgenson and Mieko Nishimizu, "Closing the Technology Gap: The United States versus Japan, with Some Inferences for Other Industrial Countries," *New International Realities*, Vol. III, No. 1 (Winter 1978), pp. 7–15.

2   *In my view, this statement is not correct. Although the level of technology present in Japanese and German industry is approaching that of U.S. industry, I do not believe that either has yet surpassed it.* —**William S. Anderson**

3   *This statement is not elaborated in the McCulloch paper and the Jorgenson-Nishimizu paper is not listed in her bibliography. Thus, more convincing proof for this statement is required before it can be accepted as part of the argument.* —**Howard W. Bell**

ability to compete with imports has been declining—which again depends in part on R&D.

For these reasons, the Committee on Changing International Realities decided to include a study of R&D in its series on the determinants of U.S. international competitiveness. Professor Rachel McCulloch of Harvard University was well-qualified to undertake this assignment and her study provides a comprehensive analysis of the main aspects of the subject. Without necessarily endorsing all of her analyses and conclusions, we are pleased to recommend that Professor McCulloch's work be published by NPA as a study signed by its author.

We wish to emphasize one of the most important implications of Professor McCulloch's study: the need for more effective government efforts to encourage R&D in the private sector and for greater government support for R&D, particularly basic research, in certain fields, such as energy and the environment.[4]

Essentially, U.S. government encouragement and support for private-sector R&D are justified because the benefits to American society as a whole clearly exceed the benefits that accrue to the business firms, universities, research institutes, and individuals carrying on the R&D.[5] The social benefits include not only those we have already noted—the raising of living standards through economic growth and the improvement of international competitiveness—but also the contributions to health and safety, national security, and the advancement of human knowledge generally. Because the organizations and individuals who conduct R&D in the private sector do not retain all of the benefits thereby created, it is in the interest of American society as a whole to provide financial support whenever they lack adequate incentives and means for carrying on their work.[6] For, if the government does not assure that reasonable incentives and assistance are available, the private sector is not likely to invest the socially optimal amount of

---

4   *The McCulloch paper makes clear that pure research in all areas should be supported by government because private firms cannot capture the benefits. However, applied R&D can and should be performed by the private sector with proper and supportive government policies.*—**Howard W. Bell**

5   *In my view, the major thrust of government encouragement and assistance should be toward policies and actions that enable the private sector to undertake and carry through research and development which inures to the benefit of the society as a whole as well as to business firms, universities, research institutes, etc.*—**Nathaniel Samuels**

6   *The Committee Statement expresses too broad a generalization on the desirability of government financial support for private-sector R&D. The statement ". . . it is in the interest of American society as a whole to provide financial support whenever they lack adequate incentives and means for carrying on their work" surely cannot be accepted without qualification of the word "whenever." Some areas of private-sector research should clearly be left to the workings of the competitive free-enterprise system for allocation of financial support. If inadequate incentives exist in the view of individual competitors, then research should not be supported. Broadly speaking, these areas involve research and development of commercial products or processes by those who desire and are able to commercialize the results.*

There are areas of research where there is a broad public benefit and where the intended result is a commercial product or process but where incentives for private R&D may be inadequate because of the cost, scope or length of the development process, or because of market uncertainties. Given these conditions, some sort of balanced private-sector/government involvement seems appropriate. Definition of the public benefit which will accrue as a result of government support should be a required discipline in the selection of such areas for government involvement.—**Jack G. Clarke**

resources in R&D. This is especially the case in such fields as the development of new energy sources and the protection of the environment, where effective R&D requires large expenditures over long periods and then substantial capital investments in order to apply the results.

For these reasons, we are concerned over the decline in government-sponsored basic research and government encouragement and support for private-sector R&D. Accordingly, we urge that existing policies with respect to tax incentives and federal funding of R&D be reassessed in the light of current and prospective needs for technological innovation to sustain U.S. economic growth and international competitiveness.[7,8,9]

---

7   *There are R&D benefits from foreign direct investments which may not appear in expenditure data. Europe and Japan benefited greatly in the 1960s and 1970s from U.S. direct foreign investment which carried with some delay the fruits of R&D undertaken here. The United States also benefited from European investments and know-how flowing here. With European and Japanese direct investments in the United States now increasing greatly, the United States will be the beneficiary of rising R&D arising in those areas. Thus, the areas for R&D expenditures can be identified but we should recognize that the benefits go beyond these areas. As international capital flows for direct investments increase, this matter gains significance.* —**Jerome Jacobson**

8   *The terms "government encouragement and support for private-sector R&D" should not, in our opinion, be read as referring primarily to government payments or tax considerations for R&D per se. Of comparable or greater importance are changes in those government controls and regulatory procedures which prevent private industry from capitalizing on successful technological innovation in a timely fashion, and discourage U.S. companies from competing in world markets.* —**James E. Lee**

9   *I agree with the point made on page 15 that the United States can derive benefits from technological advances abroad as reflected in higher quality or lower prices for imported goods. Nonetheless, I would argue that the relative technological gains of other nations in relation to the United States is a definite cause for concern—particularly in those high-technology industries such as computers and semiconductors where the United States has held a traditional comparative advantage. In these industries, companies with leading edge technologies have an inherent advantage in gaining market penetration around the world. As the gap closes, the United States stands to lose jobs and a valuable source of export revenues. For this reason, it is imperative that these U.S. industries invest the necessary resources in R&D to keep this lead from shrinking away.* —**Mark Shepherd, Jr.**

ix

EDWARD LITTLEJOHN
Vice President–Public Affairs, Pfizer, Inc.

WILLIAM J. McDONOUGH
Executive Vice President, International Banking Department, The First National Bank of Chicago

WILLIAM R. MILLER
Executive Vice President, Bristol-Myers Company

ALFRED F. MIOSSI
Executive Vice President, Continental Illinois National Bank and Trust Company of Chicago

WILLIAM G. MITCHELL
President, Central Telephone & Utilities Company

WILLIAM S. OGDEN
Executive Vice President, The Chase Manhattan Bank, N.A.

*J.S. PARKER
Vice Chairman of the Board, General Electric Company

WILLIAM R. PEARCE
Corporate Vice President, Cargill Incorporated

RALPH A. PFEIFFER, JR.
Chairman of the Board, IBM, World Trade Americas/Far East Corporation

THOMAS A. REED
Group Vice President, International Control Systems, Honeywell Incorporated

JERRY REES
Executive Vice President, National Association of Wheat Growers

*NATHANIEL SAMUELS
Chairman, Louis Dreyfus Corporation; Limited Partner, Kuhn, Loeb & Company

DANIEL I. SARGENT
General Partner, Salomon Brothers

CHARLES R. SAYRE
President and General Manager, Staple Cotton Cooperative Association

*MARK SHEPHERD, JR.
Chairman of the Board and Chief Executive Officer, Texas Instruments, Incorporated

WILLIAM F. SPENGLER
President and Chief Operating Officer, International Operations, Owens-Illinois

WALTER STERLING SURREY
Senior Partner, Surrey, Karasik and Morse

THOMAS N. URBAN
Executive Vice President, Pioneer Hi-Bred International

JOHN A. WAAGE
Vice Chairman of the Board, Manufacturers Hanover Trust Company

MARINA v. N. WHITMAN
Professor of Economics, University of Pittsburgh

KEMMONS WILSON
Chairman of the Board, Holiday Inns Incorporated

WILLIAM S. WOODSIDE
President, American Can Company

RALPH S. YOHE
Editor, *Wisconsin Agriculturist*

The opinions expressed and the recommendations presented in the Committee Statement are solely those of the individual members of the Committee on Changing International Realities whose signatures are offered hereto and do not represent the views of the National Planning Association or its staff. Committee members' agreement or disagreement with specific points of this Statement is expressed in signed footnotes.

---

*See footnotes to the Statement.

Introduction 1

The United States has long occupied a position of acknowledged world leadership in science and its industrial application. However, U.S. technological preeminence has not remained unchallenged. In 1957, the Soviet Union's Sputnik triumph shocked the nation. The fears engendered by Sputnik resulted in massive new public support for research and development (R&D) and for education of the scientists and engineers required to carry out the new programs. More recently, commercial rather than strategic aspects of competition from abroad have become the focus of concern. High-technology products continue to make a consistent and important contribution to U.S. trade performance, but significant technological gains by U.S. trading partners have aroused fears that the strength of the American economy may be in danger. Critics of U.S. policy, pointing to diminished federal support for research and development, increased restrictions on the introduction of new products, and accelerated transfers abroad of advanced American technology, have called for positive measures to maintain U.S. technological superiority.

Technology has played a major role in America's transformation from a small, primarily agricultural economy into the foremost industrial power of the modern world. The nation's development reflected the combined impact of three major forces: rapid population growth through both immigration and high rates of natural increase, territorial expansion, and steady gains in productivity associated with continuing technological innovation.[1] The role of technology in U.S. economic growth can be divided into three phases. Until well into the nineteenth century, American entrepreneurs expanded per capita output by adapting to local conditions innovations originating primarily in Europe, particularly in Great Britain. By the end of the Civil War, U.S. technology was on a par with that of Europe, and American inventors began to add in a significant way to the rapidly expanding world stock of technical knowledge. Finally, in the years following World War I, the United States gradually assumed a still more active role in the production of new knowledge, until, in the post-World War II era, the nation emerged as the unquestioned world leader in research and its industrial application.

During the 1950s and 1960s, U.S. private and government spending for R&D soared, increasing fivefold in current dollars and threefold in real terms. As a percentage of gross national product (GNP), total R&D expenditures rose from 1.5

Suggestions and comments of members of the Committee on Changing International Realities on an earlier draft of this report are gratefully acknowledged. The author is also indebted to Richard Freeman, Raymond Vernon, Zvi Griliches, E.M. Graham, and James Utterback for helpful discussions.

1   Simon Kuznets, "Two Centuries of Economic Growth: Reflections on U.S. Experience," *American Economic Review, Papers and Proceedings*, Vol. 67, No. 1 (February 1977), pp. 5–8.

TABLE 1-1

Absolute Resources Devoted to R&D in OECD Countries, 1970–71

(millions of U.S. dollars)

High		Medium		Low	
Over $25,000		$600–1,200		$300–600		$75–150		$2–35	

High $2,500–4,500		Medium		Low	

United States	27,336								

High		$600–1,200		$300–600		$75–150		$2–35	
West Germany	4,499	Canada	1,165	Sweden	538	Denmark	143	Ireland	33
Japan	4,041	Italy	929	Switzerland	486	Norway	112	Portugal	24
France	2,920	Netherlands	784	Belgium	364	Finland	91	Greece	18
United Kingdom	2,596			Australia	341	Spain	78	Iceland	3
						Austria	78		

Source: *Patterns of Resources Devoted to Research and Experimental Development in the OECD Area, 1963–1971* (Paris: Organization for Economic Cooperation and Development, 1975), pp. 9–10.

percent in 1953, peaking at 3 percent in the years 1964–67. At the same time, the share of education in national income also rose rapidly, resulting in important increases in the educational attainments of new labor force entrants and in the supply of scientists and engineers.[2] Although motivated largely by other considerations, these investments in R&D and skilled manpower were reflected in the nation's trade statistics. Until the early 1960s, the overall U.S. trade balance showed a persistent surplus, with particularly strong performance in the high-technology industries. During this period, the "technology gap" between the United States and Europe was the focus of much concern on both sides of the Atlantic. Europe saw itself falling ever further behind, eventually becoming an "economic and technological colony of the United States."[3]

Unforeseen developments led to a rapid change in this picture. As the postwar recovery proceeded, other industrialized nations were able to increase their expenditures for research and development. Over the same period, the fraction of U.S. GNP devoted to R&D leveled off and then declined. By the early 1970s, a few countries had actually surpassed the United States in total R&D expenditures as a fraction of GNP, although the absolute level of U.S. spending still dwarfs that of its commercial rivals (see Table 1–1). At the same time, increased transfers abroad of advanced technology through foreign investment and licensing by U.S. firms also helped to narrow the technology gap. The orientation of foreign R&D differed markedly from that in the United States, where defense and space exploration have constituted major priorities. Much of the activity abroad has been directed specifically toward the development of marketable products. Some countries—notably Japan—have concentrated on the adaptation of imported technologies.

By 1971–72, the overall U.S. trade balance showed a large deficit, although the high-technology industries still fared relatively well. The combined trade surplus of the technology-intensive industries rose from $7.7 billion to $11.1 billion between 1963 and 1969, while the deficit for all other manufactured goods grew from $1.0 billion to $7.5 billion. Nevertheless, the evident success of Europe and especially Japan in displacing American manufactured goods in world markets for low-technology products, an important source of concern in itself, has also led U.S. businessmen and policy makers to anticipate similar encroachments in the markets for more sophisticated products as the U.S. technological lead is gradually eliminated.[4]

---

2   Richard B. Freeman, "Investment in Human Capital and Knowledge," in *Capital for Productivity and Jobs*, Eli Shapiro and William L. White, eds. (Englewood Cliffs, N.J.: Prentice-Hall, 1977), pp. 97–98.

3   For a detailed description of this view of the technology gap, see Harvey Brooks, "Have the Circumstances that Placed the United States in the Lead in Science and Technology Changed?", in *Science Policy and Business: The Changing Relation of Europe and the United States*, David W. Ewing, ed., John Diebold Lectures, 1971 (Boston: Harvard University Graduate School of Business Administration, 1973), p. 13.

4   See, for example, Michael Boretsky, "Trends in U.S. Technology: A Political Economist's View," *American Scientist*, Vol. 63 (January-February 1975), pp. 70–82; and Robert Gilpin, *Technology, Economic Growth, and International Competitiveness*, Joint Economic Committee, U.S. Congress (Washington, D.C., 1975).

Although some observers have argued that these changes are merely the inevitable and even desirable consequences of growth and development abroad,[5] the events of the past decade raise important questions about the appropriate role of the U.S. government in fostering economic growth through policies to promote research and development. As direct federal expenditures for R&D have waned, there has been increasing attention paid to government policies which indirectly affect the profitability of industrial R&D undertaken by the private sector. Of particular interest are safety and environmental regulations, which appear to stimulate successful innovation in the private sector while slowing down the growth of productivity as conventionally measured.[6]

The purpose of this study is to shed light on a number of issues related to the appropriate role of government in fostering R&D and the possible implications of R&D policy choices for U.S. international competitiveness. Chapter 2 presents a statistical assessment of U.S. R&D activity in relation to that of other industrial nations. In Chapter 3, available evidence concerning the impact of R&D on economic performance is summarized. The current and potential role of government policy as a determinant of R&D is evaluated in Chapter 4. Conclusions emerging from the data and analysis are presented in Chapter 5.

---

5   Brooks, "Have the Circumstances that Placed the United States," pp. 16–18.

6   Thomas J. Allen et al., "Government Influence on the Process of Innovation in Europe and Japan," *Research Policy*, Vol. 7, No. 2 (April 1978), pp. 124–149. GNP statistics understate true productivity gains when these take the form of such unmarketed collective goods as improved air and water quality.

The State of U.S. Technology 2

AN OVERVIEW OF CURRENT U.S. EXPENDITURES FOR R&D

Approximately $38 billion was spent for R&D activities in the United States during 1976.[1] This figure represents an 8 percent increase over 1975 in terms of current dollars and about a 2 percent increase in constant dollars. As a share of GNP, however, R&D has continued to fall from a peak of 3.0 percent in the mid-1960s; the current share is 2.3 percent. The declining fraction of GNP devoted to R&D is attributable mainly to a reduced rate of growth of federal support, particularly in the categories of defense and space, over the past decade.

Research and development comprise a wide range of activities. Basic or pure research seeks to extend the boundaries of scientific knowledge. In the United States, more than one-half of all basic research is undertaken in colleges and universities. In applied research, known scientific principles are directed to a specific practical use. Successful applied research typically yields a new product or process which may be of potential commercial value. Development, undertaken mainly in private industry, is concerned with solving production problems and improving product design.[2] Although there is a natural progression from a scientific breakthrough to its successful commercial application, applied research or development may also stimulate advances in basic science. For example, this can happen through accumulation of new information as a product or process becomes more widely used or as a result of unforeseen technological problems encountered in translating a new product or process from the laboratory to the production line.

According to National Science Foundation estimates, $4.8 billion of the 1976 total (12 percent) was used for basic research, $8.9 billion (23 percent) for applied research, and $24 billion (64 percent) for development. However, the allocation of expenditures among the three categories is necessarily somewhat arbitrary. Furthermore, the total R&D figure is subject to two systematic biases which lead to an understatement of the resources actually used for R&D activities by smaller firms and to an overstatement for larger firms. In both cases, the source of the bias is the lack of a clear distinction between R&D, particularly development, and routine production activities such as quality control. Smaller firms with no separate R&D staff or budget allocation may nevertheless perform some R&D activities; similarly,

---

1   National Science Foundation, *Science Resource Studies Highlights*, May 21, 1976, p. 1.

2   For the definitions used in NSF surveys of R&D activity, see National Science Foundation, *National Patterns of R&D Resources, 1953–1976* (Washington, D.C., 1976), p. 17.

the special R&D departments of larger firms are likely to carry out some routine functions related to current production.

The major source of R&D support is the federal government, which currently supplies more than one-half of total funds, almost $22 billion in 1976. This support is concentrated in a few areas, particularly those in which the federal government is the major consumer. Spending for defense-related R&D in 1976 constituted about one-half of the total, with space (13 percent), health (11 percent), and energy (8 percent) the next largest allocations. Table 2–1 gives a complete breakdown of total federal support by function for fiscal year 1976. Nonfederal support for R&D was about $18 billion in 1976, with industry providing $16.6 billion, 92 percent of the total.

Most R&D is performed by private industry—about 70 percent of all R&D in 1976, 54 percent of applied research, 86 percent of development, but just 16 percent of basic research. In contrast, colleges and universities, which account for only 10 percent of all R&D, perform about 55 percent of basic research. The federal government performs 15 percent of all R&D, 16 percent of basic research, 25 percent of applied research, and 11 percent of development. The 1976 shares of total R&D effort by performing sector are shown in Table 2–2. Appendix Tables A–1 through A–3 give a complete breakdown of R&D effort by performer and source of funds.

Of industry-performed R&D, six industries—aircraft and missiles, electrical equipment and communication, machinery, chemical and allied products, motor

TABLE 2–1

U.S. Federal R&D, Fiscal Year 1976

Function	Billions of Dollars	Percent of Total
Total	$21.6	100.0%
National defense	10.6	49.2
Space	2.9	13.3
Health	2.4	10.9
Energy	1.6	7.5
Basic science	0.9	4.0
Environment	1.0	4.5
Transport and communications	0.7	3.3
Natural resources	0.5	2.3
Agriculture	0.4	1.9
Education	0.2	0.9
Income security and social services	0.2	0.7
Area and community development, housing, public services	0.1	0.6
All other programs	0.2	0.8

Details may not total because of rounding.

Source: National Science Foundation, *Science Resource Studies Highlights*, August 19, 1976, p. 3.

TABLE 2–2

Performance of Research and Development, by Sector, 1976
(percent)

	Federal Government	Industry	Colleges and Universities	Associated FFRDCs*	Other Nonprofit Institutions
Total R&D	14.7%	69.6%	9.6%	2.8%	3.3%
Basic research	15.8	16.3	54.7	7.1	6.1
Applied research	25.2	53.8	10.3	4.3	6.5
Development	10.6	85.7	0.5	1.5	1.6

*Associated Federally Funded Research and Development Centers.

Details may not total because of rounding.

Source: NSF, *National Patterns of R&D Resources, 1953–1976* (Washington, D.C., 1976), pp. 20–27.

vehicles and motor vehicle equipment, and professional and scientific instruments—accounted for 85 percent of total expenditures in 1975. Table 2–3 presents a breakdown of 1975 industrial R&D by source of support and type of research for each of the six research-intensive industries, for all other manufacturing industries, and for nonmanufacturing industries.

TABLE 2–3

U.S. Industrial R&D, 1975
(billions of dollars)

	Total	Source of Funds		Allocation of Funds		
		Federal	Company	Basic	Applied	Development
All industries	$23.5	$8.8	$14.8	$0.7	$4.4	$18.4
Aircraft and missiles	5.7	4.5	1.2	0.0	0.6	5.0
Electrical equipment and communication	5.5	2.5	3.0	0.2	0.9	4.5
Machinery	2.7	0.4	2.3	0.0	0.3	2.3
Chemicals and allied products	2.6	0.2	2.4	0.3	1.0	1.3
Motor vehicles and equipment	2.3	0.3	2.0	0.0	0.1	2.2
Professional and scientific instruments	1.0	0.2	0.9	0.0	0.1	0.9
Other manufacturing	2.9	0.2	2.7	0.2	1.1	1.7
Nonmanufacturing	0.8	0.5	0.3	0.0	0.3	0.5

Source: NSF, *Science Resource Studies Highlights*, October 27, 1976, p. 2.

HOW DOES THE UNITED STATES COMPARE WITH OTHER COUNTRIES?

The large absolute size of the U.S. economy and the nation's commanding lead in most areas of science and technology complicate the problem of evaluating current American R&D activities in relation to those of other countries. In many fields, other nations have allocated a major part of R&D funds for adaptation to their own commercial and strategic requirements of the fruits of past U.S. R&D efforts. Even when national defense considerations have prompted the United States to limit access of other nations to its advanced technology, scientists abroad have been able to duplicate U.S. results at a small fraction of the original cost.

Although the United States has also derived considerable benefits from imported scientific and technological knowledge, its relative position has meant that this source of advances could be of only secondary importance. However, as other industrialized nations are able to narrow the technology gap, the United States will benefit accordingly. Furthermore, because labor costs in Europe and Japan are now approaching, and in some cases exceeding, those in the United States, and because all industrialized nations are likely to face secularly rising prices for many raw materials in the future, innovations originating abroad will be of increasing interest to U.S. producers and consumers.

The United States is still spending more on research and development than the combined total for all other Organization for Economic Cooperation and Development (OECD) countries. Nevertheless, critics argue that the United States is "falling behind," relative to recent efforts of other industrialized nations and to its own past performance. The National Science Board, in a recent evaluation of U.S. R&D activities, made a number of specific international comparisons, each of which sheds light on some aspect of U.S. performance in relation to that of other nations.[3] Although each individual comparison has serious defects as a measure of

TABLE 2-4

R&D as a Percent of GNP, 1963 and 1973

	1963	1973
United States	2.9%	2.4%
Canada	0.9	0.9
France	1.5	1.7
West Germany	1.4	2.4
United Kingdom	2.3	1.9
Japan	1.2	1.9
USSR	2.2	3.1

Sources: National Science Board, *Science Indicators, 1974* (Washington, D.C., 1975), p. 154 (except Canada, United Kingdom); *U.S. International Economic Report of the President, 1976*, p. 119 (Canada, United Kingdom).

---

3   National Science Board, *Science Indicators, 1974* (Washington, D.C., 1975), pp. 2–3.

TABLE 2–5

Scientists and Engineers Engaged in R&D, 1963–73
(per 10,000 population)

	1963	1964	1969	1971	1973
United States	—	24.7	27.5	25.6	24.9
USSR	18.8	20.3	29.1	32.6	37.2
Japan	12.0	—	16.9	18.9	—
West Germany	—	5.7	12.5	14.9	17.8
France	6.7	—	10.9	11.1	—

Source: *Science Indicators, 1974*, p. 155.

U.S. technological effort and capability, a consistent overall picture can be obtained from available data.

(1) *R&D as a Fraction of GNP.* For the United States, this proportion has been falling steadily since the late 1960s. In contrast, the fraction of GNP devoted to R&D has been rising in West Germany, Japan and the USSR. Table 2–4 shows percentages for 1963 and 1973. (Appendix Table A–4 presents the figures for other years.) The downward trend for the United States continued in 1975, with R&D expenditures dropping to 2.3 percent of GNP.

(2) *Scientists and Engineers.* After 1969, the number of scientists and engineers engaged in R&D as a fraction of total population fell for the United States in response to sharp reductions in federal support of basic science.[4] Elsewhere this proportion continued to rise. However, the U.S. fraction remains higher than that of its major commercial competitors. Table 2–5 compares the United States with other major R&D-performing countries.

(3) *Defense Spending.* R&D allocated to national defense functions is likely to have a relatively minor impact on productivity growth and to produce fewer commercially viable innovations than R&D directed toward other objectives.[5] The fraction of government R&D devoted to national defense fell for all major R&D-performing countries between 1961 and the early 1970s. However, the United States allocated the largest fraction of total R&D resources to national defense of any OECD nation.[6] Table 2–6 compares the United States with four major commer-

---

4   Richard B. Freeman, "Investment in Human Capital and Knowledge," in *Capital for Productivity and Jobs*, Eli Shapiro and William L. White, eds. (Englewood Cliffs, N.J.: Prentice-Hall, 1977), p. 98.

5   Harvey Brooks has argued that U.S. defense and space research diverted scientific manpower and venture capital from civilian R&D. See Harvey Brooks, "Have the Circumstances that Placed the United States in the Lead in Science and Technology Changed?", in *Science Policy and Business: The Changing Relation of Europe and the United States*, David W. Ewing, ed., John Diebold Lectures, 1971 (Boston: Harvard University Graduate School of Business Administration, 1973), pp. 24–25.

6   Although the USSR does not publish statistics on its allocation of R&D resources, the fraction of total R&D devoted to defense is believed to exceed that of the United States.

TABLE 2-6

**Percent of Government R&D Allocated to
National Defense, 1961–62 and 1971–72**

	1961–62	1971–72
United States	71%	53%
United Kingdom	65	44[a]
France	44[b]	28[c]
West Germany	22[b]	15[d]
Japan	4	2[e]

[a]1972–73; [b]1961; [c]1972; [d]1971; [e]1970–71.

Source: *Science Indicators, 1974*, p. 156.

cial competitors. (See Table 2–1 for more detail on the distribution of government R&D expenditures.)

(4) *Output of Scientific Literature.* Publications in technical journals provide a rough index of research output, i.e., production of new knowledge. American researchers contributed the largest share of the scientific literature published in 1973 of any major R&D-performing country in all fields except chemistry and mathematics, where the USSR had the highest share. Although the relative and absolute position of the United States changed little in this respect between 1965 and 1973, there was some evidence of a slight decline in both 1972 and 1973 in the fields of chemistry, engineering and physics, possibly a consequence of reduced funding for basic research in these areas.[7]

(5) *Literature Citations.* Citation indexes compare a nation's actual share of citations worldwide with its share of total publications in that field and thus provide a measure of the quality of research output. By this measure, the United States leads or is tied for first place in every field assessed, with the largest leads in physics and the earth and space sciences.[8] Language of publication may play an important role in determining the frequency of citations. This appears to be the case for the USSR, which ranks last in five out of six fields. Nevertheless, Japan ranks just below the United States and the United Kingdom in medicine, biology, and the earth and space sciences.

(6) *Nobel Prizes.* An award to scientists responsible for major advances in their fields, the Nobel prize is a measure of national research performance in the basic sciences. U.S. scientists have received a greater share of Nobel prizes in the sciences than scientists of any other nation. The U.S. share fell, however, between 1951–60 and 1961–70 in all fields. Furthermore, when shares are adjusted for total

---

7   *Science Indicators, 1974*, p. 9.

8   Ibid., p. 13.

population, the United States falls behind the United Kingdom for the decades since World War II.[9]

   (7) *Patent Balance.* Patent statistics provide a crude measure of "inventiveness." The number of foreign patents granted to U.S. individuals less the number of U.S. patents granted to foreign nationals, the U.S. "patent balance" was positive but declining between 1966 and 1973. This observed decline reflects at least in part an increase in the degree of integration of the world economy, an explanation which is supported by corresponding declines in the patent balances of other nations. As of 1973, the United States had a positive balance with all countries except West Germany and the USSR.[10] Patent statistics should be interpreted with care because patented inventions vary considerably in their scientific and economic significance. Furthermore, some important innovations are not patented. Also, legal as well as technological factors play an important role in determining patent practices and changes in them. Table 2–7 shows the patent balances of the United States with other R&D-performing nations.[11]

   (8) *Major Innovations.* A recent study of major innovations between 1953 and 1973 showed the United States to have originated a majority.[12] However, the

TABLE 2–7

U.S. Patent Balance,[1] 1966–73

	1966	1970	1973
Worldwide	36,066	33,697	25,306
Canada	15,676	17,598	11,619
West Germany	−248	−1,152	−639
Japan	3,561	2,149	546
United Kingdom	11,440	9,776	8,866
Other EEC countries[2]	5,700	5,743	4,914
USSR	−63	−17	−177

[1]Foreign patents to U.S. nationals less U.S. patents to foreign nationals.

[2]Excluding France.

Source: *Science Indicators, 1974,* Tables 1–10 and 1–11, p. 164

9   Ibid., pp. 13–15. In 1976, however, all Nobel prizes awarded went to U.S. scientists.

10   Ibid., pp. 16–17.

11   If patent statistics are to be used for international comparisons of commercial inventiveness, a more meaningful measure would be relative penetration of third-country markets. This method compares national performance under a common set of legal and economic conditions.

12   "Indicators of International Trends in Technological Innovation," a report prepared for the National Science Foundation by Stephen Feinman and William Fuentevilla, Gellman Research Associates, Inc., April 1976.

TABLE 2–8

**U.S. Net Payments from Patents, Manufacturing Rights
and Licenses, 1960–74**
(millions of U.S. dollars)

	1960	1967	1974
Worldwide	$210	$289	$601
Western Europe	105	97	200
Japan	48	91	241
Developing nations	25	48	91
Other	31	54	69

Details may not total because of rounding.

Source: *Science Indicators, 1974*, p. 167.

U.S. share fell from a high of 80 percent in the late 1950s to 55–60 percent toward the end of the period. The United Kingdom was second, followed by West Germany and Japan, both of which increased their shares between 1960 and 1973. Relative to its total number of innovations, the United Kingdom led the United States in the proportion classed as "major technical advance" or "radical break-through."[13]

(9) *Sale of Technical Information.* Receipts from sales of technology to other countries provide a measure of a nation's accumulated stock of useful knowledge. Purchases of technical information from other countries indicate the nation's ability to make use of such information. This in turn depends both on the relative advancement of domestic technology and on the existence of the infrastructure needed to adapt imported products and processes to local requirements. The United States had a positive and increasing balance of payments from the sale of patents, licenses and manufacturing rights for the period 1960–74. Japan was a major purchaser of U.S. technology, a fact which may indicate Japan's relative sophistication in science. Table 2–8 shows the U.S. balance arising from sale of "know-how."

(10) *Productivity.* Gross domestic product (GDP) per employed civilian de-pends upon many aspects of economic structure, including the capital-labor ratio, state of technology, and labor skills. This measure of productivity remains higher in the United States than in other major R&D-performing countries, although other nations have experienced more rapid increases than the United States[14] (see Table 2–9).

(11) *Balance of Trade.* The "product cycle" hypothesis identifies differences in technology across countries as a principal determinant of trading patterns. The

---

13   *Science Indicators, 1974*, p. 20.

14   Ibid., pp. 22–24.

TABLE 2–9

GDP per Employed Civilian, 1960–74
(U.S. = 100)

	1960	1967	1974
United States	100	100	100
France	55	63	81
West Germany	52	56	74
Japan	24	32	57
United Kingdom	51	49	56

Source: *Science Indicators, 1974*, p. 168.

United States, with its technological lead in most areas, has experienced a consistently large and positive balance of trade in high-technology products. A recent empirical study of the export performance of U.S. manufacturing industries concluded that up to three-fourths of total variation in industries' export performance is associated with differences in research intensity alone.[15]

U.S. performance in high-technology industries appeared to weaken between 1968 and 1972, but the high-technology surplus has been growing again since 1973.[16] In contrast, the balance of trade for other manufactured goods has been

TABLE 2–10

U.S. Trade Balance for Selected Commodities, 1960–75
(billions of dollars)

	1960	1964	1968	1972	1975
Aircraft and parts	$1.0	$0.8	$2.0	$2.5	$ 5.7
Computers and parts	0.0	0.2	0.5	1.2	2.1
Other nonelectric machinery	2.6	3.4	3.6	4.3	12.5
Basic chemicals and compounds	0.1	0.5	0.7	0.7	1.6
Motor vehicles and parts	0.6	1.1	−0.6	−3.5	−0.6
Steel products	0.2	−0.1	−1.4	−1.9	−1.7
Consumer electronics	−0.1	−0.2	−0.6	−1.7	−1.6
Textiles, clothing and footwear	−0.4	−0.5	−1.5	−3.3	−3.0
All manufactures	6.2	8.0	4.5	−2.4	22.4

Source: *U.S. International Economic Report of the President, 1976*, Table 27, p. 151.

15   Thomas C. Lowinger, "The Technology Factor and the Export Performance of U.S. Manufacturing Industries," *Economic Inquiry*, Vol. 13, No. 2 (June 1975), pp. 221–236.

16   *U.S. International Economic Report of the President, 1977*, pp. 120–124.

negative in recent years. Table 2–10 shows movements in the U.S. trade balance for selected commodities over the period from 1960 to 1975. While high-technology industries generated increasingly large surpluses, trade deficits steadily grew in motor vehicles, steel, consumer electronics, and textiles, clothing and footwear—a trend reversed only after devaluation of the dollar. Following a large overall balance-of-trade surplus for 1975, U.S. trade performance again began to sag. Although the unprecedented recent deficits are partly a consequence of accelerated oil imports, a number of manufacturing industries also encountered renewed problems with competing imports; the federal government has received urgent appeals for import relief from manufacturers of shoes, color television sets and steel, among others. In 1975, negative balances on U.S. trade with Japan and Organization of Petroleum Exporting Countries (OPEC) were offset by positive balances on trade with the European Economic Community, the communist countries and non-OPEC developing nations, producing an overall trade surplus for the year. Table 2–11 shows these components.

Obviously, no single measure allows a meaningful assessment of the state of U.S. technological capabilities. In interpreting the statistics presented above, it is useful to keep in mind that some measures, such as the balance of payments on foreign sales of technological information or the index of scientific citations, depend upon the accumulated national *stock* of scientific and technical knowledge, while other measures, especially the proportion of GNP devoted to R&D, reflect current *additions* to that stock of knowledge. The U.S.'s showing is most clearly superior to that of other nations in terms of measures which depend upon the accumulated stock and not as strong in measures of current additions to that stock. Thus, the composite picture which emerges shows the nation still preeminent in science and technology, but with its competitors closing the lead in some important respects.

TABLE 2–11

U.S. Balance of Trade by Region, 1975
(billions of dollars, f.a.s.*)

	Balance
Worldwide	$11.1
Canada	0.0
EEC	6.3
Japan	−1.7
Less-developed countries except OPEC	6.3
OPEC members	−6.3
Communist countries	2.2

*Free alongside ship.

Source: *U.S. International Economic Report of the President, 1976*, Table 22, p. 149, and Tables 28–33, pp. 152–153.

Whether the relative gains of other nations should be a source of concern to the United States is subject to debate. First, it is a nation's absolute rather than its relative technological gains which are the primary long-run determinant of its economic growth and welfare. Furthermore, in many cases, the United States can also derive benefits from technological advances abroad as these become reflected in lower prices of imported goods. But, rapid changes in international comparative advantage associated with technological advances abroad have posed a serious internal adjustment problem for the United States. It is the problem of adjustment to changed economic conditions, rather than foreign technological progress, which is likely to cause welfare losses at home. Larger R&D investments in the United States may mitigate problems of adjustment by delaying the time at which a particular industry begins to lose ground to its foreign competitors and in some cases may even allow the industry to remain competitive indefinitely. For this reason, R&D assistance may appropriately be regarded as a possible domestic policy tool for minimizing the private and social costs associated with adjustment to changed international competitive conditions.[17]

---

17   U.S. shoe manufacturers injured by competition from imports have received federal technical and financial assistance in modernizing their operations under an experimental U.S. Department of Commerce program.

R&D in Relation to Productivity Growth and Competitiveness 3

EFFECTS OF R&D ON PRODUCTIVITY

R&D is economic activity specifically designed to promote technological innovation. This effort may be directed toward any of the successive stages in the transformation of technical knowledge into a usable product or process. Systematic study of the relationship between R&D investments and resulting measurable outputs is complicated by the long and unpredictable lag between discovery of the basic scientific or technological information which makes an innovation possible and the large-scale use of the resulting innovation. The length of this lag is determined by economic as well as technological factors. The feasibility of a new process may be of scientific but not of commercial importance at one set of input prices, but highly profitable at other prices, as in the case of solar energy. Similarly, labor-saving appliances will be of commercial interest only if an adequately large high-wage market exists. However, the purely technological problems which must be surmounted in taking a product or process from the laboratory or workshop to the stage of mass production and distribution may in some instances be more difficult and time-consuming than the initial discovery on which the innovation is based, as in the frequently cited case of penicillin.

Despite the formidable conceptual and empirical problems entailed in quantifying the contribution of R&D to economic growth and increased productivity, a large volume of studies has led economists to virtual agreement on two basic points. The first is that improvements in technology have been a crucial determinant of economic growth in the United States and other industrialized countries. The second point is that investment in research by the private sector, and, in some instances, by government, has had a rate of return as high as and sometimes higher than other types of investment (physical capital, education).[1]

Studies evaluating the economic contribution of improved technology fall into four categories, which take as their focus individual innovations, firms, industries, or the economy as a whole. Studies which calculate rates of return for significant individual innovations relate the value of new or improved products and processes to the full costs of innovation, including associated "dead-end" research. In this category are studies by Griliches for hybrid corn, Peterson for poultry research,

---

[1] As long as all investments can be viewed as entailing many independent projects, no allowance for risk need be included in this comparison. With a small number of large projects, however, returns to R&D or other investments should be adjusted for risk. This would be the case, for example, with a massive program to develop nuclear fusion as an energy source.

TABLE 3–1

Rates of Return on R&D for Individual Innovations

Study	Coverage	Internal Rate of Return (percent)
Griliches (1958)*	Hybrid corn	37%
Peterson (1967)	Poultry breeding	33
Weisbrod (1971)	Polio vaccine	9–13

*Years refer to publication; see Bibliography.

Source: Richard B. Freeman, "Investment in Human Capital and Knowledge," in *Capital for Productivity and Jobs*, Eli Shapiro and William L. White, eds. (Englewood Cliffs, N.J.: Prentice-Hall, 1977), pp. 112–113.

and Weisbrod for polio vaccine. Their findings are summarized in Table 3–1.[2] Studies of individual innovations allow a precision in measuring direct benefits and costs which is not possible at a more aggregate level and thus provide considerable insight into the innovation process. However, their results cannot easily be generalized to predict relationships between R&D and aggregate economic growth because unsuccessful innovations are not investigated.

An alternative approach, more likely to capture the cost of all R&D effort whether or not it yields some useful innovation, relates cumulated R&D expenditures to productivity growth in the individual firm performing the R&D. Results of four such studies are summarized in Table 3–2. These results must be interpreted with caution because R&D activities may have important spillover effects on the productivity of other firms in the industry. Furthermore, new or improved products or lower prices in one industry will raise productivity in other industries downstream, even if the latter engage in little R&D activity of their own.

A third group of studies links productivity growth across industries to cumulated investment in R&D. Results of four studies of this type are summarized in Table 3–3. Although this type of analysis does capture cross-firm productivity effects, it does not reveal productivity gains to one industry resulting from improved technology in another.

In the three groups of studies described above, R&D effort enters explicitly into the analysis. A fourth approach, usually used to investigate economic growth at the aggregate level, is the "residual" or "growth accounting" method. Growth in aggregate output is related to the growth of factor inputs; that part of output growth not accounted for by growth in factor inputs is labeled "productivity increase."

---

2   Table 3–1 shows *internal* rates of return, which depend upon the distribution over time of estimated costs and benefits of a particular R&D project. Tables 3–2 and 3–3 show perpetual rates of return, the calculated annual yield to the firm or industry from an additional dollar invested in R&D.

TABLE 3–2

Rates of Return on R&D for Individual Firms

Study	Coverage	Perpetual Rate of Return (percent)
Minasian (1969)[a]	8 chemical firms	48–54%
Mansfield (1965)	10 petroleum and chemical firms	40–60 (petroleum)
		7–30 (chemical)
Baily (1972)	6 pharmaceutical firms	25–35[b]
Griliches (1976)	883 large firms	17

[a]Years refer to publication; see Bibliography.

[b]Internal rate of return.

Source: Freeman, "Investment in Human Capital," pp. 112–113.

Work in this area has concentrated on development of more satisfactory output and input measures. An early study by Solow found that growth in capital and labor accounted for only about 13 percent of U.S. growth over the period 1909–49.[3] This left a staggering 87 percent of growth to be explained by "technical change," that is, *shifts* in the production function relating inputs and outputs. Of course, the residual of unexplained output increase captured many effects omitted from Solow's analysis. In addition to improved technology, the residual could be due in part to economies of scale, improved managerial techniques, favorable

TABLE 3–3

Rates of Return on R&D for Manufacturing Industries

Study	Coverage	Perpetual Rate of Return (percent)
Terleckyj (1960)*	20 industries, 1919–53	50%
Mansfield (1968)	10 industries, 1946–62	20–62
Griliches (1973)	85 industries, 1958–63	40
Terleckyj (1974)	20 industries, 1948–66	30 (private)
		0 (government-funded)

*Years refer to publication; see Bibliography.

Source: Freeman, "Investment in Human Capital," pp. 112–113.

---

3   Robert Solow, "Technical Change and the Aggregate Production Function," *Review of Economics and Statistics*, Vol. 39, No. 3 (August 1957), pp. 312–320.

changes in the social and political environment, and so on. Probably more important, the residual also reflects deficiencies in the measurement of inputs.

The first refinements in the measure of inputs concentrated on labor skills. Over the period analyzed by Solow, the American labor force changed in "quality" as a result of increased education. A study by Denison, in which labor inputs were adjusted for increased quality, reduced growth of outputs unexplained by growth of inputs to only one-third of the total.[4] Later work by Jorgenson and Griliches[5] also adjusted capital inputs for improved quality and yielded a still smaller residual.[6]

A recent study by Christensen, Cummings and Jorgenson has compared the sources of post-World War II economic growth for nine countries.[7] For the period 1960–73, the highest-output growth rates, for Japan (10.9 percent) and Korea (9.8 percent), are associated with the highest growth rates of real capital input (11.5 and 7.3 percent, respectively) as well as of total factor productivity (4.5 and 4.0 percent, respectively). Korea also had the highest growth rate of real labor input (5.0 percent), with both hours worked and quality of hours worked increasing rapidly in comparison with other nations.

The growth accounting framework yields no satisfactory measure of the value of R&D in stimulating economic growth. As mentioned earlier, the residual may capture numerous influences apart from improved technology. Furthermore, even that part of the residual which is attributable to advances in technical knowledge need not reflect results of R&D. Denison has estimated that only about one-fifth of U.S. aggregate productivity improvement is the result of organized research activities in the United States.[8] For countries with smaller R&D investments, the figure is probably still lower. On the other hand, the growth accounting framework "credits" to real factor growth what may be the ultimate consequences of increased technological knowledge attributable to R&D. Investments in physical and human capital depend upon rates of return, and improved technology probably stimulates both types of investments. Any advance in knowledge which is useful only when "embodied" in improved capital goods will show up in the growth accounts as the contribution of input growth rather than increased factor productivity.

---

4   Edward F. Denison, *Sources of Economic Growth in the United States and the Alternatives Before Us*, Supplementary Paper 13 (Washington, D.C.: Committee for Economic Development, 1962).

5   Dale W. Jorgenson and Zvi Griliches, "The Explanation of Productivity Change," *Review of Economic Studies*, Vol. 34, No. 3 (July 1967), pp. 249–284.

6   The Jorgenson and Griliches corrections gave rise to a heated methodological debate with Denison. An exchange of views was reprinted in Dale W. Jorgenson, Zvi Griliches and Edward F. Denison, "The Measurement of Productivity," *Survey of Current Business*, Vol. 52, No. 5, Part II (May 1972), pp. 31–111.

7   Laurits R. Christensen, Dianne Cummings and Dale W. Jorgenson, "Economic Growth, 1947–1973: An International Comparison," Harvard Institute of Economic Research Discussion Paper No. 521 (December 1976). Forthcoming in *New Developments in Productivity Measurement*, J. W. Kendrick and B. Vaccara, eds., Studies in Income and Wealth, Vol. 41, National Bureau of Economic Research.

8   Charles T. Stewart, "A Summary of the State-of-the Art on the Relationship Between R&D and Economic Growth/Productivity," in *Research and Development and Economic Growth/Productivity* (Washington, D.C.: National Science Foundation, 1972), pp. 16–17.

The relationship of aggregate R&D effort to aggregate economic growth is further complicated by the heterogeneity of research effort. Expenditures for development can be expected to have a more immediate impact on economic growth than expenditures supporting basic research. Furthermore, the gains from the latter are likely to be readily available to other nations. Also, R&D expenditures related to defense and space, a large fraction of total government-supported R&D in the United States, may have little or no effect on productivity growth.[9]

R&D AS A DETERMINANT OF TRADE

Previous sections have reviewed empirical evidence on the relationship between U.S. R&D and productivity growth and on the continuing strength of U.S. trade performance in the high-technology sectors of manufacturing. This section examines channels through which R&D may influence the international competitiveness of U.S. products.

The ability of a U.S. firm to export its products depends upon the combined impact of many economic considerations: (1) dollar costs of labor, capital and other inputs; (2) factor productivity; (3) exchange rates, tariffs and quotas; (4) terms of delivery, insurance and credit; (5) product characteristics; and (6) seller reputation, service facilities and warranties.

The first four items listed together determine *delivered cost* in terms of foreign currency of the product. However, these considerations are not independent. More productive labor and capital command a higher market price. Conversely, high factor costs may provide an incentive for undertaking R&D projects designed to cut costs. High dollar costs due to inflation at home are likely to force the dollar down in value relative to other currencies. A government that is reluctant to allow its currency to depreciate may instead subsidize credit for foreign purchasers.

Delivered cost is only one dimension of competitiveness. The fifth and sixth considerations listed are sources of *differentiation* of the product from its potential competitors, features which may compensate for a relatively high delivered cost. Of course, even when a product is unique, delivered cost will enter into its export performance by determining total world demand. Also, high production costs or trade barriers induce some U.S. firms to set up production facilities abroad rather than serving foreign markets through exports only.

Cost Competitiveness. The United States has been a "high-wage" economy for many years. However, as Table 3–4 indicates, wages in other industrial countries are rising rapidly. Average labor costs in Canada and Sweden now exceed those in the United States, and West German wages are now close to U.S. levels. It should be noted that changes shown reflect both rises in local money compensation and exchange-rate realignments. Also, these aggregate statistics make no correction for differences in levels of skills, education or experience.

International wage differences are far from uniform across industries. As Table 3–5 indicates, the United States has the highest wages of any country in primary

---

9   Japan, the country which has experienced the highest productivity growth rate, has allocated the smallest share of its government support to defense, space and nuclear R&D.

TABLE 3-4

Hourly Compensation of Production Workers in Manufacturing, 1960-76
(U.S. dollars)

	1960	1970	1975	1976
United States	$2.66	$4.20	$6.33	$6.90
Canada	2.13	3.46	6.24	7.39
France	0.83	1.74	4.50	4.59
Italy	0.63	1.75	4.44	4.27
Japan	0.26	0.99	3.05	3.26
Sweden	1.20	2.96	7.36	8.50
United Kingdom	0.82	1.46	3.26	3.05
West Germany	0.83	2.32	6.33	6.70

Note: 1975 figures are midyear; 1976 figures are midyear estimates; no adjustments have been made for the value of fringe benefits.

Source: *U.S. International Economic Report of the President*, 1977, p. 99.

metals and motor vehicles. However, it is not labor cost per hour but labor cost per unit of output which is relevant in determining international competitiveness. From 1960 to 1976, productivity (output per hour) rose more slowly in the United States than in other industrialized countries, perhaps a reflection of the high level already reached at the beginning of the period. But, U.S. wages also rose more slowly than elsewhere, so that unit labor costs grew less rapidly in the United States than in Canada, Europe or Japan.[10]

As long as productivity gains are not fully reflected in higher wages, additional R&D could increase the cost-competitiveness of some U.S. manufactures in dollar terms. However, it is important to recognize that the very success of such a strategy is likely to induce changes throughout the economy which *reduce* cost competitiveness of other industries through increased factor costs and exchange-rate appreciation. This process is discussed in greater detail below.

For at least some low-skill industries, the U.S. cost disadvantage may be too great to be overcome through feasible R&D-induced productivity improvements. As Table 3-5 shows, wage rates in textiles, footwear and apparel are relatively low in all the industrial countries. However, labor costs are far lower still in less-developed nations such as Taiwan, Hong Kong and Brazil, which have gained a rapidly increasing share of the world market for these products despite formidable barriers to trade.

Product Competitiveness. Some U.S. products, usually new products and often from the high-technology industries, have no close competitors in foreign markets. The United States, with its giant (in absolute terms) research establishment and abundant supply (both in absolute and relative terms) of skilled labor,

---

10   *U.S. International Economic Report of the President, 1977*, Table 13, p. 145.

TABLE 3-5

Hourly Compensation of Production Workers by Selected Industry, 1976*

(U.S. dollars)

	Textiles	Footwear	Apparel	Chemicals and Chemical Products	Primary Metals	Electrical Equipment	Non-electrical Machinery	Motor Vehicles
United States	$4.55	$4.10	$4.15	$7.80	$9.95	$6.50	$7.55	$10.75
Canada	5.30	4.55	4.60	7.75	8.90	6.75	8.30	9.35
France	3.80	3.60	3.35	5.50	5.60	4.40	4.90	4.80
Italy	3.50	3.20	2.75	5.05	5.60	4.55	4.60	5.10
Japan	2.35	2.25	1.70	4.65	4.85	3.00	3.70	3.85
Sweden	7.60	7.25	6.95	8.30	9.40	7.85	8.35	8.80
United Kingdom	2.60	2.70	1.90	3.50	3.70	2.90	3.15	3.65
West Germany	5.50	4.80	4.95	7.60	7.30	6.25	7.05	8.60

*Midyear estimates.

Source: *U.S. International Economic Report of the President, 1977*, pp. 100–101.

scientists and engineers, has enjoyed a temporary world monopoly position for many unique products, only to have that advantage eroded through imitation or diffusion of the required technology.

Until the 1950s, economists took for granted that the United States, clearly a capital-abundant economy, had a comparative advantage in the production of capital-intensive goods—as predicted by the well-accepted factor-proportions theory of trade-flow determination.[11] The finding by Leontief that U.S. imports are more capital-intensive than U.S. exports stimulated a wave of new theoretical and empirical investigations into the determinants of world trade flows.[12] These studies highlighted the roles played by skills or "human capital," on the one hand, and R&D, on the other hand; in empirical investigations, these considerations are difficult to separate because the industries with high R&D intensity (a high ratio of R&D expenditures to total sales) are also ones with a high proportion of skilled workers.[13] Work by Keesing[14] and others confirmed that U.S. export performance was strongest in industries employing high proportions of skilled workers. Thus, the factor-proportions theory could be reinterpreted as predicting that the skilled-labor-abundant United States has its comparative advantage in the production of skilled-labor-intensive products.

A related response to the Leontief "paradox" came in the dynamic trade product cycle theory, which interpreted much of U.S. trade as reflecting the first stage of the adjustment process following a successful innovation. According to the product cycle theory, as described by Vernon[15] and others, a new product actually changes in its characteristics over time, becoming increasingly standardized and thus amenable to production by relatively less-skilled workers. In the early stages of commercial exploitation, rapid interaction between market and producer is advantageous as alternative product characteristics are being explored; reliability is often low, so that access to service facilities may be crucial; the initial innovation and production process is likely to require considerable inputs of skilled labor. As standardization proceeds and the market for a new product expands in response to falling prices, the need for skilled labor is greatly reduced, and cost competitiveness begins to exert a decisive influence on location

11   For a review of alternative trade theories and empirical evidence, see Robert M. Stern, "Testing Trade Theories," in *International Trade and Finance: Frontiers for Research*, Peter B. Kenen, ed. (Cambridge, Eng.: Cambridge University Press, 1975), pp. 3–49.

12   Wassily Leontief, "Domestic Production and Foreign Trade: The American Capital Position Re-Examined," *Economia Internazionale*, Vol. 7, No. 1 (February 1954), pp. 3–32.

13   G.C. Hufbauer, "The Impact of National Characteristics and Technology on the Commodity Composition of Trade in Manufactured Goods," in *The Technology Factor in International Trade*, Raymond Vernon, ed. (New York: Columbia University Press, 1970), pp. 145–231.

14   Donald B. Keesing, "Labor Skills and Comparative Advantage," *American Economic Review, Papers and Proceedings*, Vol. 56, No. 2 (May 1966), pp. 249–258; and idem, "The Impact of Research and Development on United States Trade," *Journal of Political Economy*, Vol. 75, No. 1 (February 1967), pp. 38–48.

15   Raymond Vernon, "International Investment and International Trade in the Product Cycle," *Quarterly Journal of Economics*, Vol. 80, No. 2 (May 1966), pp. 190–207.

decision. This will be true particularly when an innovation has been emulated successfully by other producers. The product cycle hypothesis has been well documented for many new products, including synthetic fibers, drugs and consumer electronics. Nevertheless, the product cycle is more a suggestive scenario than a complete theory because it does not predict the length of successive phases in the cycle.

Because the innovating country gradually loses its competitiveness for a given product as the cycle proceeds, some attention has been given to measures which would slow down the cycle, delaying the shift of production to lower cost locations. Among measures suggested are legal restraints on the rights of innovating firms to exploit their unique technology through direct foreign investment or licensing of foreign production. For example, the Burke-Hartke bill, endorsed by the AFL-CIO, would have allowed the President to prohibit the holder of a U.S. patent from manufacturing the product abroad or licensing foreign production if, in the President's judgment, this prohibition would contribute to increased employment in the United States. The effects of implementing such measures could, however, be very different from those intended. Most U.S. firms characterize their foreign investments as defensive, arguing that the markets served would otherwise be lost to European or Japanese firms taking advantage of favorable cost conditions abroad. Furthermore, the proposals tend to overlook the dynamic character of the product cycle. Although it would perhaps be possible to delay the shift abroad of production of particular goods, such restraints lower the profitability of innovative activity. This in turn could reduce future investments in R&D, diminishing the flow of new products with which the cycle commences. Thus, measures which stimulate the innovation process, rather than retarding the diffusion process, are more likely to yield long-run benefits.

GENERAL EQUILIBRIUM EFFECTS OF EXPANDED R&D

The previous section detailed ways in which R&D contributes to an industry's international cost or product competitiveness. However, each industry is connected to the rest of the economy in a number of ways. Factor and product market linkages are one important source of interaction. The behavior of exchange rates and endogenous elements of U.S. and foreign commercial policy is another. The technologically based advantage of affected firms will show up in the form of increased foreign sales and possibly also as decreased penetration of the domestic market by imports. These are, however, sectoral effects. The effects of expanded R&D for the welfare of the nation in general and for the structure of U.S. trade in particular must be viewed in terms of the total rather than the partial effect, taking into account important spillover effects which influence the costs, growth rates and international competitiveness of other industries. These interconnections are illustrated in Figure 1.

A first major interindustry effect comes in the form of superior inputs available to industries which buy from those developing new or improved products. Improved products sold by one industry can show up as cost reductions for others, with superior intermediate goods and capital goods making production of existing products less expensive and sometimes also facilitating development of further

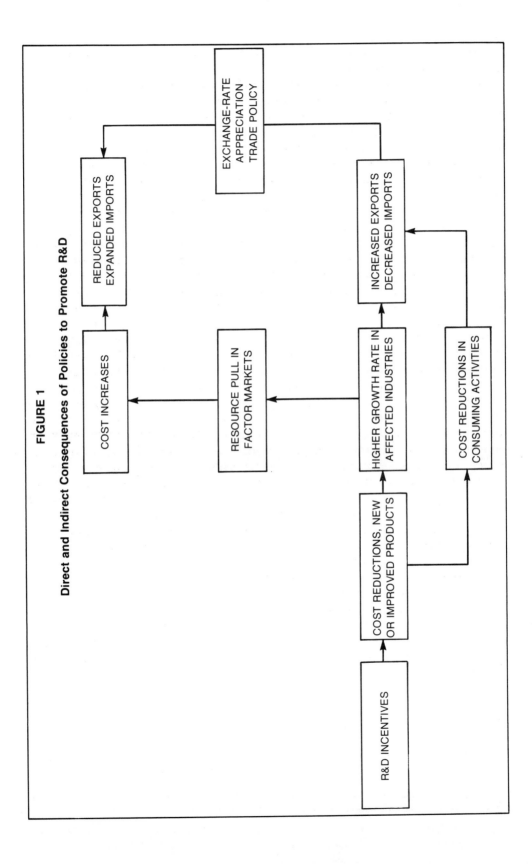

FIGURE 1

Direct and Indirect Consequences of Policies to Promote R&D

new products. Thus, successful innovation can induce a secondary wave of benefits in downstream industries.

A second source of interaction comes about because all U.S. industries are tapping an interconnected market for productive factors. Even if incentives provided are not specific to particular sectors of the economy, differences in the profitability of making R&D investments are likely to imply that more activity will be generated in relatively "new" industries where the rate of return on such activity at the margin has not yet been forced down by past innovation and imitation. Thus, the high-technology industries will probably have the largest induced responses. The innovating industries will be induced to expand as profitable new products and processes are generated. New products and lower (quality-adjusted) costs will allow these industries to increase their market shares, both domestically and abroad. The necessary expansion of these industries will draw capital and labor out of other parts of the economy and may also lead to an expansion of overall employment. Mobile factors drawn into the expanding industries will tend to receive higher financial rewards, as a bonus for their willingness to relocate, either geographically, by industry, or in terms of new skills required. In the United States, the high-technology industries have grown about twice as fast as the low-technology industries and have created new jobs about five times as fast.[16]

As the technologically progressive industries expand, drawing in capital and labor, the prices of some productive factors will rise, resulting in higher costs for other industries requiring these inputs. This pull creates a natural and desirable incentive for stagnant sectors of the economy to contract, but it may also cause some dislocations as declining industries become unable to compete in international markets. Older industries with slower growth rates are likely to suffer from the exit of capital, entrepreneurial talent, and younger and more skilled workers, as these resources are drawn into the more profitable expanding new sectors.[17]

A third spillover comes through balance-of-trade and exchange-rate consequences of successful innovation. With a flexible rate system, the expansion of exports by the technologically progressive sectors will induce an exchange-rate appreciation. This makes other sectors of the economy less cost competitive even if they have been unaffected by factor market spillovers. When exchange rates are fixed or "managed," improved trade performance in some sectors may influence overall commercial policy by weakening the case for protectionist or mercantilist options. Thus, vigorous trade performance by some sectors may lead to a more liberal trade stance and less likelihood that broad-based balance-of-trade measures such as import deposit requirements, import surtaxes or disguised export inducements will be adopted or retained. Although this is highly desirable for the economy as a whole, it may exacerbate the already considerable adjustment problems of stagnant industries.

---

16   "The Vital Need for Technology and Jobs," a speech by Dr. Thomas A. Vanderslice, Vice President, General Electric Company, presented to the Executives' Club of Chicago, November 5, 1976.

17   Japan has actually hastened this process by exporting the capital equipment of labor-intensive industries to lower-wage developing nations such as Korea as local labor costs rise.

Government Policy toward R&D 4

THE NEED FOR GOVERNMENT ACTION
TO PROMOTE R&D

Promotion of research and development has come to be recognized as a legitimate and important function of the federal government. Approximately $23.5 billion was allocated for support of R&D activities during the 1977 fiscal year, around 5 percent of the total federal budget. However, these figures underestimate the total level of resources channeled to R&D activities by the federal government because they do not include the indirect costs (including foregone federal revenues) of policies intended to encourage innovative activity, such as accelerated depreciation of new equipment.

Federal support for R&D can be justified in one of three ways, not mutually exclusive. First, the lion's share of federally supported R&D is in areas of public-sector functions, particularly national defense and space (see Table 2–1, page 6). Because the government is the major and often the only purchaser of the outputs of these sectors of the economy, it must also undertake the support of research and development aimed at producing improvements in these areas.[1] Civilian "spin-offs" from defense and space R&D programs reduce their net cost to the nation; however, spin-offs do not in themselves constitute an economic justification of the programs, as the same level of resources directed toward civilian objectives would yield a higher level of civilian benefits.[2]

A second justification for federal R&D support pertains in particular to basic science research and to a lesser extent to applied R&D activities. Most advances in basic knowledge have little or no immediate market value, as they are useful mainly in the production of further knowledge rather than salable goods and services. Also, new knowledge is a "public good" in the sense that it cannot be used up and therefore yields the greatest social benefits when made freely available to all

---

1   Edwin Mansfield, "Federal Support of R&D Activities in the Private Sector," in *Priorities and Efficiency in Federal Research and Development*, Joint Economic Committee, U.S. Congress (Washington, D.C., 1976), pp. 88–89.

2   Harvey Brooks has argued that "the concentration of R&D in a narrow range of sophisticated technologies for defense and space, in the United States, has diverted innovative talent and energy as well as venture capital away from civilian industry and from public needs other than defense and national prestige . . . the spin-off from space-defense spending, especially R&D spending, has been largely a negative factor in the economy." See Harvey Brooks, "Have the Circumstances that Placed the United States in the Lead in Science and Technology Changed?", in *Science Policy and Business: The Changing Relation of Europe and the United States*, David W. Ewing, ed., John Diebold Lectures, 1971 (Boston: Harvard University Graduate School of Business Administration, 1973), pp. 14–15.

potential users. Private innovators will not have the incentive to produce as much new knowledge as is desirable from the national (or world) point of view; furthermore, benefits to private innovators rely mainly on maintaining exclusive access to the newly created knowledge, rather than making it freely available to potential users. For these reasons, direct government support of basic science research is generally acknowledged to be necessary and desirable. However, the political process in the United States is such as to make funds more readily available for project-oriented research than for basic research. Thus, much fundamental biological and medical research has been supported by funds allocated for the "War on Cancer."

In the case of applied research and development in the civilian sector, resulting innovation typically has direct commercial usefulness. In this case, there is a tradeoff between the *static* benefits from making any given knowledge freely available to all potential users and the *dynamic* benefits from ensuring a steady flow of new innovations by allowing the innovator to exploit commercially the benefits from exclusive access. The patent systems in use in the United States and other countries represent a practical compromise between these considerations for those types of commercially useful knowledge which are subject to control through patents. However, the total social gains from innovation of this type may still far exceed the benefits which can be captured through commercial exploitation. This may be true because of spillover effects, as when the new or improved products of one industry lower costs in another, or because innovation is highly risky and requires immense capital outlays long in advance of expected benefits, as in the area of energy. To the extent that social benefits from innovation generally exceed private benefits, so that the private sector will systematically underinvest in applied R&D, there is a case for government policies which raise the overall private return through measures such as favorable tax treatment or subsidies. These measures would be relatively neutral with respect to different industries and types of innovation,[3] but they would raise the *average* return to innovative activity of all types, thus leaving to individual firms the choice of areas deemed most promising. However, if particular areas such as energy R&D are considered to have especially high risk or prohibitive capital requirements, there may be a case for special measures to raise the private return to innovation in these industries *relative to* the average rate for all industries.

A number of industrialized countries have provided R&D incentives specifically designed to establish or maintain "international competitiveness" of particular industries or of manufacturing as a whole. Although this has not been an explicit policy in the United States, it has been a frequent justification offered in support of individual proposals and of R&D incentives generally. In this connection, two issues should be raised. First, as with any investment decision, the case for R&D requires not merely evidence of a positive effect, but of a (social) rate of return higher than that for alternative uses of the resources. Second, as discussed earlier, R&D incentives which improve the competitiveness of some sectors may have indirect consequences which lead to a deterioration of the trade position of other industries.

---

3   This does not mean, of course, that R&D would increase by the same percentage or dollar amount in each industry.

R&D aimed at public-sector functions or basic science presents a special problem of resource allocation which arises to a much smaller degree in promotion of commercially useful innovation. In the latter, market demand provides a guide to resource allocation. In the former, it is far more difficult to attach dollar values to expected outcomes of alternative allocations; appropriate allowances for risk and for comparison of present and future benefits may be highly subjective. Allocation of R&D support for public-sector functions relies upon a largely bureaucratic decision-making process with important inputs from the potential users of innovations. Many U.S. Department of Defense contractors also initiate their own R&D activities to develop new products or processes geared to perceived future public-sector requirements. The treatment of costs incurred for such contractor-initiated R&D has been an area of contention.[4]

In basic science research, most funds are allocated by a "peer review" process. This has the advantage of engaging the judgment of those likely to be most knowledgeable in any given field, but it may also systematically discriminate against heterodox approaches. Another frequent criticism of current practice is that awards depend as much upon the professional reputation of the principal investigator, based mainly on past research performance, as on the scientific merit of the proposed work. Although this procedure has the virtue of introducing important additional information into the decision process, it may reinforce the position of established scientists rather than encourage the work of innovative but less well-known researchers. A study recently undertaken by the National Academy of Sciences seeks to review past research support decisions made by the National Science Foundation to determine the extent to which the principal investigator's identity affects evaluation of proposed research.[5]

THE POLICY SPECTRUM

Government policies exert a profound influence upon the level and effectiveness of R&D activity. Some government measures are specifically intended to stimulate innovation. Many more measures are concerned with other aspects of national economic performance, but they nonetheless have an important effect—positive or negative—on innovative activity, almost certainly greater in the aggregate than that of policies bearing specifically on R&D. In assessing the policy spectrum available for influencing R&D, this section examines not only those measures intended to encourage R&D, but also those policies affecting R&D which have other underlying objectives. Each channel of influence has its distinctive

---

4   Some members of Congress view the expenses of independent research and development (IR&D) initiated by federal contractors as a logical target for cost-reduction efforts. The contractors say that these industry efforts are a legitimate cost of doing business and must be recovered in the price of goods sold—including those sold to the government. For the industry case, see *Technical Papers on Independent Research and Development and Bid and Proposal Efforts* (Washington, D.C.: Tri-Association Ad Hoc Committee on IR&D and B&P, March 1974).

5   For some preliminary findings on the peer review process, see Stephen Cole, Leonard C. Rubin and Johnathan R. Cole, "Peer Review and the Support of Science," *Scientific American*, Vol. 237, No. 4 (October 1977), pp. 34–41.

advantages and drawbacks. The more direct the means of encouraging R&D, the more control the government can exercise over the nature of the work undertaken.[6] However, the more direct the stimulus, the smaller the role left to be played by market-generated incentives to maximize the economic returns from innovative activity, through choice of the most promising prospects and through control of costs. When policies are geared to objectives other than influencing R&D, it may be difficult or impossible to tailor provisions in such a way as to increase positive incentives to innovation or to mitigate the effects of negative ones.

Government actions which affect R&D are classified under the following five headings which distinguish among policies by the degree of support provided and the extent to which the effect on R&D is a primary motivation.[7] This classification indicates the very broad range of government policies which can be expected to influence national R&D activity.

Direct Performance. This category, which accounts for about 15 percent of U.S. R&D, includes all those projects directly undertaken by government agencies in government-controlled facilities.

Direct Support. A number of policies provide full or partial funding of research carried out in universities or other nonprofit facilities and in industry. About 35 percent of R&D funding in the United States comes in this form. Possible types of direct support include subsidies, joint government-industry ventures, special loan funds to finance innovations and their commercial application, and government procurement policies.

Primary Incentives. The purpose of these policies is to affect the incentives for innovative activity, but they entail no direct budgetary allocation of funds. Special tax treatment of R&D (including accelerated depreciation of capital equipment embodying new technology) is one major incentive program of this type.[8] Also included is national policy regarding the terms and lifetime of patent rights and government support of the education of scientists and engineers through fellowship programs and grants to universities.

Secondary Incentives. Many policies affect the cost of or returns to innovative activity even though this is not their immediate goal. Of recent concern are policies to control foreign direct investment and technology transfer; although maintenance of employment and wages in import-impacted industries provides the primary motivation for action of this type, the implied restrictions on the economic

---

6   The degree of control achieved by "tying" of R&D funds to particular priority areas may be rather small, however, especially when R&D outcomes are difficult to specify in advance, as with most basic research.

7   For the United States, level of direct support appears to be highest for those areas of R&D which are viewed as most essential to national welfare, for example, defense. In Japan, direct controls rather than financial support play an important role, and level of direct support is not likely to be a good indicator of national priority.

8   It should be noted that the normal treatment of most R&D costs is already more favorable than that of other capital expenditures. See Mansfield, "Federal Support of R&D," p. 91.

usefulness of new technological knowledge lower the expected returns to firms engaging in innovative activity. Similarly, strict health and safety standards for new products, although intended primarily to protect consumers from product hazards, raise the cost of commercial introduction of new products and hence lower the expected returns from innovation.[9] In the same way, massive R&D efforts in the areas of defense and space probably have raised R&D costs to other industries through their impact on the salaries of scientists and engineers.[10]

Incidental Incentives. Policies intended to improve the overall efficiency of the market mechanism through government regulation of economic activity may have indirect and sometimes unintended effects on incentives for innovation. Rate of return and price regulation of public utilities affect the return to innovation and to adoption of new technology as it becomes available. Similarly, when significant economies of scale exist in R&D or in the commercial adoption of new products and processes, successful antitrust action may lower the profitability of innovation if it reduces the size of firms in a given industry.[11] Of course, in neither case does the existence of negative incidental incentives necessarily argue against retention of present policies. However, it is essential that government agencies charged with implementing these policies be fully aware of such possible effects. (Although they are not directly the consequence of government action, union work rules delaying the introduction of new techniques or equipment are also likely to reduce incentives for innovation.)

Government regulation may also provide positive incentives for innovation. In recent years, pollution and automobile safety regulations have been responsible for greatly increased industry-initiated R&D in these areas. A recent study of industrial innovation in Europe and Japan found that government regulatory restraints were frequently associated with successful innovations.[12] Minimum wage laws (or effective action by unions to raise wages) are likely to increase the profitability of labor-saving inventions. Price-support programs for agricultural products or natural resources increase the profitability of developing synthetic substitutes. However, it should be emphasized that even if a particular policy results in increased R&D activity, it need not be beneficial to the economy as a whole.

Secondary and incidental incentives may be of great interest in assessing current prospects for increasing the level of R&D undertaken in the private sector. The adverse consequences for innovative activity of policies the primary objective of which lies in other areas are often of great weight in determining the overall

---

9   This case has been put most strongly by manufacturers of pharmaceuticals, but probably applies to a number of other industries as well.

10   Brooks, "Have the Circumstances that Placed the United States," pp. 14–15.

11   See Robert Gilpin, *Technology, Economic Growth, and International Competitiveness*, Joint Economic Committee, U.S. Congress (Washington, D.C., 1975), pp. 41–44; and Mansfield, "Federal Support of R&D," pp. 110–112.

12   Thomas J. Allen et al., "Government Influence on the Process of Innovation in Europe and Japan," *Research Policy*, Vol. 7, No. 2 (April 1978), pp. 124–149.

structure of incentives facing industrial innovators. At a time of public disillusion-ment with federally funded research, it may be difficult or impossible to increase budgetary allocations required for expansion of government performance or direct support. Similarly, additional tax incentives may be hard to implement. Changes in indirect incentives, especially those which now work to lower the returns to innovative activity in the private sector, may offer an important and often unrecognized option for promoting an expansion of R&D in the United States.

R&D PROGRAMS OF U.S. TRADING PARTNERS

Most industrialized nations have adopted programs intended to promote R&D as a means of increasing competitiveness in world markets. These programs may operate directly, by generating commercially viable innovations and stimulating their diffusion, or indirectly, by strengthening the nation's overall technological capabilities. In addition to competitiveness in international trade, national prestige and defense considerations provide further motivation for adoption of policies to stimulate R&D. Table 4–1 reviews programs and tax incentives in use by major U.S. trading partners.

The actual impact of government policies to promote R&D depends critically on the general environment in which economic activity is conducted. Specifically, the overall state of technological advancement of the economy, market structure and other governmental regulatory behavior may be expected to play key roles in determining the outcome of a specific policy measure. A major study of foreign experience recently completed by the Massachusetts Institute of Technology Center for Policy Alternatives under a grant from the National Science Foundation has provided some important evidence concerning government influence on innovation.[13] The M.I.T. study sought to identify country-specific and industry-specific factors influencing the process of innovation in five countries—France, Germany, the Netherlands, the United Kingdom, and Japan—and five industries—computers, consumer electronics, textiles, industrial chemicals, and automobiles. Information from 59 firms supplied researchers with a total sample of 164 cases of industrial innovation. On the basis of interviews with managers, 66 cases were judged successful by the companies' own criteria and 51 as unsuccess-ful; 47 were still in progress. According to the managers, government involvement was present in almost one-half of the projects. The most frequent forms of influence were R&D cost reduction, environmental or safety regulation, and policies to facilitate access by firms to new technology. Government involvement was most frequently perceived by managers as a negative rather than a positive influence on project performance.

Except in the case of environmental and safety regulations, the rate of project success did not appear to depend upon government involvement. However, if firms correctly assessed the probability of success prior to embarking on a project,

---

13   Center for Policy Alternatives, *National Support for Science and Technology: An Evaluation of Foreign Experiences* (Cambridge: M.I.T., 1976); and Allen et al., "Government Influence," report results from this study concerning the effects of government policy on the innovation process.

TABLE 4-1

Government R&D Programs of U.S. Trading Partners

	Program	Tax Treatment
WEST GERMANY	No elaborate plan for stimulating industrial R&D although government supports various scientific organizations.	Tax allowances for R&D expenses incurred by corporations and individual inventors.
UNITED KINGDOM	Preproduction order support program to accelerate the use of technologically advanced capital goods (machine tools) in industry. Financial support for R&D in industry program for further application of technology in small firms and research associations. Launching aid program provides interest-free loans for development of civilian aircraft and engines. National Research Development Corporation, a public corporation, develops and exploits inventions from publicly financed research.	No special tax treatment.
FRANCE	Concerted actions program combines the efforts of universities and government laboratories for basic research on projects for industry. Aid to development program provides subsidies for development of new products for export or for import substitution. Letter of agreement guarantees the difference between actual sales and break-even point, if sales are lower, to companies developing high-priority R&D projects.	R&D expenses deductible up to total technical research profits. Accelerated depreciation on the first 50 percent of cost of R&D facilities.
JAPAN	Japanese Research Development Corporation, a quasi-public corporation, provides 60 to 80 percent of development costs for high-risk R&D products with good potential for industrial use. National R&D program fully subsidizes high-priority national projects to develop new technologies. Joint government/private-sector projects in atomic energy, space and ocean development areas.	Accelerated depreciation on the first one-third of acquisition cost of capital goods related to use of new technology. Deduction of 25 percent on R&D expenses.
CANADA	Program to enhance productivity provides up to 50 percent of feasibility study costs to determine whether new technologies can improve productivity. National Research Council, a public corporation, does basic research on R&D projects for industrial use. R&D consortia in private sector encouraged in the interest of spreading risk, pooling resources and avoiding duplication.	Industrial Research and Development Incentives Act (IRDIA) designed to relieve industry of some of the financial burden of R&D effort. Provides up to 25 percent of capital expenditures for R&D. IRDIA grants nontaxable.

Source: *U.S. International Economic Report of the President, 1975,* pp. 106-107.

those undertaken without (positive) government involvement were expected to have a higher success rate. Thus, government influence may in fact raise the probability of success for projects which would not otherwise have been undertaken.

A surprising finding of the study is that environmental and safety regulations, perceived by managers as a negative influence on the innovation process, operated far more frequently in successful than in unsuccessful projects. Two possible explanations of this phenomenon may be offered. Regulatory requirements are set with existing technological capabilities in mind. Furthermore, innovative activity stimulated by these regulations is likely to produce suitably modified versions of products or processes with previously established market value.

Important cross-industry and cross-country differences in patterns of perceived government involvement were evident in the sample. Although R&D funding was the prevalent form of action in computers and electronics, technical assistance was most important in textiles, and the negative effects of environmental and safety regulations were most significant for chemicals and automobiles. It is interesting that although four of the countries had measures explicitly designed to encourage development of the computer industry, government's perceived impact was slight in three countries and negative in the fourth, the United Kingdom.[14] Across countries, German policies appeared to affect mainly the earliest stages of product development, while the role of government at late stages in the innovation process seemed most important for the U.K. firms. Although government involvement appeared to be greatest in France and least in the Netherlands, the difference was not statistically significant.

Because the total number of cases studied is small, the results of the M.I.T. study should be interpreted with caution. However, the study appears to confirm the importance of the "secondary" and "incidental" incentives affecting R&D discussed above. As Allen et al. note in their discussion of the findings, the most striking aspect is the failure to identify any systematic influence of government support on project performance.[15] The observed gap between intended and actual effects of government action underscores the difficulty of tailoring national R&D programs to the achievement of specific policy objectives.

U.S. TECHNOLOGY ENHANCEMENT PROGRAMS

The United States has done less than its major trading partners to encourage R&D in the private sector, perhaps because just a decade ago the technology gap between the United States and the rest of the industrialized world was still viewed by some as a permanent feature of the international economy. However, the rapidity with which the gap narrowed alarmed many Americans, and in recent years new federal programs have been established to promote R&D in various ways. These programs are described briefly on the following page.

---

14   However, managers in exporting firms may have played down the positive role of government to avoid any suggestion of export subsidization.

15   Allen et al., "Government Influence."

National R&D Assessment Program. This program, established in 1972 to study the role of science and technology in the U.S. economy, operates within the Directorate of Scientific, Technological, and International Affairs of the National Science Foundation. Work conducted in the program is intended to guide policy makers by identifying policy issues and clarifying the consequences of alternative options. Policy studies are carried out both by professional staff within NSF and by university-affiliated researchers.[16]

Experimental Research and Development Incentives Program. The main objective of this NSF program is to develop, on an experimental basis, measures to reduce barriers to innovation. Among the experiments carried out have been making federal laboratory facilities available to public contractors for performance validation, making university research capabilities available to industries currently performing little R&D, and establishing community programs to develop entrepreneurial ability.[17]

Experimental Technology Incentives Program. This program, within the National Bureau of Standards, was created "to find ways to stimulate R&D and the application of R&D results." ETIP focuses upon government procurement and regulatory policies as tools for promoting innovation to increase productivity growth in public-sector functions.[18]

Technical Assistance to Import-injured Industries. The Trade Act of 1974 provides for trade adjustment assistance to firms injured by increased imports. Under the program, loans, loan guarantees and technical assistance are made available to import-impacted firms, thus reducing the cost to firms of adopting newer technologies. A U.S. Department of Commerce program to aid U.S. shoe manufacturers in modernizing their operations was instituted in 1977. If the program succeeds in helping U.S. shoe producers to become more competitive, a similar approach is likely to be extended to other U.S. industries injured by imports.

INTERNATIONAL COMPARISONS[19]

Appendix Table A–4 compares government-funded R&D as a share of GNP for countries with major R&D programs. Of the group with high levels of absolute resources devoted to R&D (the United States, the United Kingdom, West Ger-

---

16   *Technological Innovation and Federal Government Policy* (Washington, D.C.: National Science Foundation, January 1976), p. 1–3.

17   Mansfield, "Federal Support of R&D," p. 106.

18   Ibid. Also see Gilpin, *Technology, Economic Growth, and International Competitiveness*. A progress report on ETIP is reprinted as an appendix to Gilpin's study.

19   A recent OECD study has surveyed changing national R&D priorities for 1961–72. See *Changing Priorities for Government R&D* (Paris: Organization for Economic Cooperation and Development, 1975).

many, France, and Japan), the United States has the highest proportion of government-funded R&D. Japan is distinguished by its very low share of government funding. However, this by no means implies a small governmental role in promoting technological progress. Rather, it reflects the close cooperation between government and industry characteristic of the Japanese economy.

Table 4–2 shows shares of total government-funded R&D by function. The U.S. defense share, 52.6 percent in 1971–72, dwarfs other U.S. programs as well as the defense R&D efforts of the other nations. France and the United Kingdom have followed the U.S. lead in this respect, with large fractions of total government R&D support allocated to national defense. It is notable that neither Japan nor Germany—the two nations most frequently cited for the rapid advance of their

TABLE 4–2

Percent Distribution of Government R&D Support, by Function

	U.S. 1971–72	Canada 1972–73	France 1972	U.K. 1972–73	W. Germany 1971	Japan 1969–70
National security and big science						
Defense	52.6%	8.6%	27.8%	42.8%	15.0%	2.2%
Civil space	18.1	1.5	6.7	1.9	6.6	0.7
Civil nuclear	5.1	13.8	14.6	8.9	15.6	7.5
Economic development						
Agriculture, forestry and fishing	2.0	18.6	3.5	4.8	2.1	14.0
Mining, manufactures	3.7	29.5	13.3	16.1	10.6	7.0
Economic services	2.3	5.4	3.3	2.3	0.8	2.2
Community services						
Health	8.5	10.8	1.8	5.0	2.5	1.8
Pollution	0.9	1.9	—	—	0.4	—
Public welfare	2.4	—	0.7	0.3	⎰ 1.3	1.7
Other community services	1.2	0.8	0.8	0.8	⎱	0.7
Advancement of science						
Advancement of research	2.9	7.6	14.6	5.8	7.6	0.2
University (general funds)	—	—	10.9	9.7	32.9	61.2
Other						
Developing countries	0.2	1.1	1.6	0.5	⎰ 4.6	—
Miscellaneous	0.0	0.3	0.2	1.1	⎱	0.8

Details may not total 100 percent because of rounding.

Source: *Changing Priorities for Government R&D* (Paris: Organization for Economic Cooperation and Development, 1975), pp. 309–322.

technological capabilities—has had the "benefits" of defense R&D spin-offs. Another noteworthy similarity between these two nations is the large share of total government support which goes to university general-research funds.

National technological capability may be defined in terms of two overlapping functions: *creation* of new technology and *adaptation and diffusion* of new technology.[20] During the 1950s and 1960s, a major focus of European and Japanese R&D was the importation and application of innovations originating mainly in the United States—an emphasis endorsed by U.S. policy makers for strategic reasons. In Europe, much of the imported technology came in the form of direct investment by American firms. Japan, however, discouraged direct investment; licensing served as the major vehicle for technology transfers.

Recent patent statistics suggest that Europe and Japan may have gained considerable ground in the capacity to create new technology as well as to adapt innovations of foreign origin. The share of U.S. patents granted to foreign residents has doubled between 1960 and 1975 (to 35 percent in 1975), with Germany and Japan accounting for the largest numbers. However, these shifts may to some extent merely reflect increasing international economic integration over the period.

---

20   Keith Pavitt, " 'International' Technology and the U.S. Economy: Is There a Problem?", in *The Effect of International Technology Transfers on U.S. Economy* (Washington, D.C.: National Science Foundation, July 1974).

Conclusions 5

Previous chapters have analyzed a number of aspects of R&D activity and its relation to innovation, productivity growth and international competitiveness. Because the basic determinants of innovative success are still open to question, the available evidence cannot be used to support strong specific recommendations for new U.S. programs. However, a number of important conclusions and general recommendations emerge from the data and analysis.

(1) In absolute terms, total U.S. R&D expenditures continue to dwarf those of other nations and indeed of all other OECD nations combined. However, the allocation of R&D effort across nations differs markedly. Germany and Japan, the nations which have made the most progress in terms of international competitiveness of industrial exports, have spent far less on defense and big science, but more on general university research support than the United States, the United Kingdom or France. Japan, which has made the most rapid productivity gains of any industrialized nation in the post-World War II period, has a very small program of direct R&D support, but uses close industry-government ties to achieve a high level of innovative activity in industry.

(2) Although the United States has slackened the pace of its R&D efforts relative to other nations, it remains preeminent by most measures of technological capacity. However, continuation of present trends is likely to produce a further narrowing of the technology gap between the United States and other countries. In the past, Europe and Japan have relied to a large extent on adaptation of imported technology, often of U.S. origin, for productivity growth. But, Germany and Japan are now rapidly approaching the United States in their capacity to create new civilian technologies.

(3) Absolute rather than relative technological advancement is the primary long-run determinant of national welfare. Narrowing of the technology gap between the United States and its trading partners can yield benefits to the United States through lower import prices and expanded opportunities to adapt technological innovations originating abroad. However, because U.S. competitiveness in international markets is currently strongest for new and unique products, weakest for standardized products in which high labor costs outweigh the U.S. factor productivity advantage, further technological gains abroad are likely to exacerbate the trade adjustment problems of some U.S. industries. For these industries, federal R&D support may be an appropriate part of industry trade adjustment assistance programs.

(4) Proposed policies to restrict the transfer abroad of advanced U.S. technology and thus slow down the product cycle could be counterproductive in their effects on U.S. competitiveness. If prevented from establishing foreign subsidiaries or licensing foreign production, U.S. firms currently serving foreign

markets through exports may lose these markets to rivals abroad with lower costs. Furthermore, restrictions on the use abroad of new technology are likely to reduce the profits of innovating U.S. firms, thus deterring future R&D investments by these firms.

(5) Because knowledge is a public good, governmental support for R&D, particularly in the area of basic science, is required to ensure a socially adequate rate of production. The case for specific policies to foster industrial R&D (applied research and development) is weaker because industrial innovators are able to capture more of the gains from R&D investments. Also, recent case studies of actual industry experience abroad suggest that government policies to promote R&D are perceived by firms to have little or no effect on performance.

(6) The apparent gap between intended and actual effects of government action underscores the difficulty of tailoring national R&D programs to the achievement of specific policy objectives. Recent evidence has established that government policies with *other* primary objectives may be crucial in determining the level and success of industrial R&D activity. This appears to be particularly true in the case of safety and environmental regulation. The role of these policies requires more careful attention from policy makers and administrators.

APPENDIX TABLES

TABLE A–1

Allocation of Funds for Basic Research, by Source and Performer, 1976
(millions of U.S. dollars)

Source of Funds	Total, All Performers	Research Performed by			
		Federal Government	Industry	Colleges and Universities	Other Nonprofit Organizations
Total	$4,750	$750	$775	$2,600	$625
Federal government	3,210	750	170	1,825	465
Industry	715	—	605	75	35
Colleges and universities	525	—	—	525	—
Other nonprofit organizations	300	—	—	175	125

Source: National Science Foundation, *National Patterns of R&D Resources, 1953–1976* (Washington, D.C., 1976), p. 23.

TABLE A–2

Allocation of Funds for Applied Research, by Source and Performer, 1976
(millions of U.S. dollars)

Source of Funds	Total, All Performers	Research Performed by			
		Federal Government	Industry	Colleges and Universities	Other Nonprofit Organizations
Total	$8,925	$2,250	$4,800	$915	$960
Federal government	4,825	2,250	1,250	545	780
Industry	3,645	—	3,550	35	60
Colleges and universities	250	—	—	250	—
Other nonprofit organizations	205	—	—	85	120

Source: NSF, *National Patterns of R&D Resources*, p. 25.

TABLE A–3

Allocation of Funds for Development, by Source and Performer, 1976
(millions of U.S. dollars)

| Source of Funds | Total, All Performers | Research Performed by | | | |
		Federal Government	Industry	Colleges and Universities	Other Nonprofit Organizations
Total	$24,415	$2,600	$20,925	$145	$745
Federal government	12,095	2,600	8,780	80	635
Industry	12,190	—	12,145	10	35
Colleges and universities	40	—	—	40	—
Other nonprofit organizations	90	—	—	15	75

Source: NSF, *National Patterns of R&D Resources*, p. 27.

TABLE A–4

R&D Expenditures as a Share of GNP, by Country, 1963–73

| | Government-funded | | | | | Industry-funded | | | | |
	1963	1967	1969	1971	1973	1963	1967	1969	1971	1973
United States	1.9	2.0	1.7	1.5	1.3	0.8	1.1	1.1	1.0	1.0
United Kingdom	1.3	1.3	1.2	—	1.1	1.0	1.1	1.1	—	0.8
Japan	0.4	0.0	0.4	0.4	0.6	0.9	0.9	1.1	1.2	1.4
West Germany	0.6	0.7	0.7	0.9	0.9	0.8	1.0	1.0	1.2	1.4
France	1.0	1.4	1.2	1.1	1.0	0.7	0.8	0.7	0.7	0.6
Canada	0.6	0.9	0.9	0.8	0.6	0.3	0.4	0.4	0.4	0.3

Sources: For 1963–71—*Patterns of Resources Devoted to Research and Experimental Development in the OECD Area, 1963–71* (Paris: OECD, 1975), Table V, p. 93. For 1973—*U.S. International Economic Report of the President, 1976.*

BIBLIOGRAPHY

Allen, Thomas J., James M. Utterback, Marvin A. Sirbu, Nicholas A. Ashford, and J. Herbert Hollomon. 1978. "Government Influence on the Process of Innovation in Europe and Japan." *Research Policy*. Vol. 7, No. 2 (April), pp. 124–149.

Baily, M.N. 1972. "Research and Development Costs and Returns: The U.S. Pharmaceutical Industry." *Journal of Political Economy*. Vol. 80, No. 1 (January/February), pp. 70–85.

Boretsky, Michael. 1975. "Trends in U.S. Technology: A Political Economist's View." *American Scientist*. Vol. 63 (January-February), pp. 70–82.

Brooks, Harvey. 1974. "Science and the Future of Economic Growth." *Journal of the Electrochemical Society*. Vol. 121, No. 2 (February), pp. 35c–42c.

—————. 1973. "Have the Circumstances that Placed the United States in the Lead in Science and Technology Changed?" In *Science Policy and Business: The Changing Relation of Europe and the United States*, David W. Ewing, ed. John Diebold Lectures, 1971. Boston: Harvard University Graduate School of Business Administration, pp. 13–32.

Center for Policy Alternatives. 1976. *National Support for Science and Technology: An Evaluation of Foreign Experiences*. Cambridge, Mass.: Massachusetts Institute of Technology.

Christensen, Laurits R., Dianne Cummings and Dale W. Jorgenson. 1977. "Economic Growth, 1947–1973: An International Comparison." Harvard Institute of Economic Research Discussion Paper No. 521. In *New Developments in Productivity Measurement*, J.W. Kendrick and B. Vaccara, eds. Studies in Income and Wealth, Vol. 41. National Bureau of Economic Research, forthcoming.

Cole, Stephen, Leonard C. Rubin and Jonathan R. Cole. 1977. "Peer Review and the Support of Science." *Scientific American*. Vol. 237, No. 4 (October), pp. 34–41.

Denison, Edward F. 1972. *Accounting for U.S. Economic Growth*. Washington, D.C.: Brookings Institution.

—————. 1962. *Sources of Economic Growth in the United States and the Alternatives Before Us*. Supplementary Paper 13. Washington, D.C.: Committee for Economic Development.

Freeman, Richard B. 1977. "Investment in Human Capital and Knowledge." In *Capital for Productivity and Jobs*, Eli Shapiro and William L. White, eds. Englewood Cliffs, N.J.: Prentice-Hall, pp. 96–123.

Gellman Research Associates. 1976. "Indicators of International Trends in Technological Innovation." Prepared by Stephen Feinman and William Fuentevilla for the National Science Foundation. Washington, D.C.

Gilpin, Robert. 1975. *Technology, Economic Growth, and International Competitiveness*. Prepared for the Joint Economic Committee, U.S. Congress. Washington, D.C.

Griliches, Zvi. 1976. "Returns to Research and Development in the Private Sector." NBER Conference on Research in Income and Wealth. In *New Developments in Productivity Measurement*, J.W. Kendrick and B. Vaccara, eds. Studies in Income and Wealth, Vol. 41. National Bureau of Economic Research, forthcoming.

_____. 1973. "Research Expenditures and Growth Accounting." In *Science and Technology in Economic Growth*, B. Williams, ed. New York: Macmillan, pp. 59–95.

_____. 1958. "Research Costs and Social Returns: Hybrid Corn and Related Innovations." *Journal of Political Economy*. Vol. 66, No. 5 (October), pp. 419–431.

Hufbauer, G.C. 1970. "The Impact of National Characteristics and Technology on the Commodity Composition of Trade in Manufactured Goods." In *The Technology Factor in International Trade*, Raymond Vernon, ed. New York: Columbia University Press, 145–231.

Johnson, Harry G. 1975. *Technology and Economic Interdependence*. London: Macmillan & Co.

Jorgenson, Dale W., and Zvi Griliches. 1967. "The Explanation of Productivity Change." *Review of Economic Studies*. Vol. 34, No. 3 (July), pp. 249–284.

Jorgenson, Dale W., Zvi Griliches and Edward F. Denison. 1972. "The Measurement of Productivity." *Survey of Current Business*. Vol. 52, No. 5, Part II (May), pp. 31–111.

Keesing, Donald B. 1967. "The Impact of Research and Development on United States Trade." *Journal of Political Economy*. Vol. 75, No. 1. (February), pp. 38–48.

_____. 1966. "Labor Skills and Comparative Advantage." *American Economic Review, Papers and Proceedings*. Vol. 56, No. 2 (May), pp. 249–258.

Kuznets, Simon. 1977. "Two Centuries of Economic Growth: Reflections on U.S. Experience." *American Economic Review, Papers and Proceedings*. Vol. 67, No. 1 (February), pp. 1–14.

Leontief, Wassily. 1954. "Domestic Production and Foreign Trade: The American Capital Position Re-Examined." *Economia Internazionale*. Vol. 7, No. 1 (February), pp. 3–32.

Lowinger, Thomas C. 1975. "The Technology Factor and the Export Performance of U.S. Manufacturing Industries." *Economic Inquiry*. Vol. 13, No. 2 (June), pp. 221–236.

Mansfield, Edwin. 1976. "Federal Support of R&D Activities in the Private Sector." In *Priorities and Efficiency in Federal Research and Development*. Joint Economic Committee, U.S. Congress, Washington, D.C.

_____. 1965. "Rates of Return from Industrial Research and Development." *American Economic Review, Papers and Proceedings*. Vol. 55, No. 2 (May), pp. 310–322.

Minasian, Jora R. 1969. "Research and Development, Production Functions, and Rates of Return." *American Economic Review, Papers and Proceedings*. Vol. 59, No. 2 (May), pp. 80–85.

National Science Board. 1975. *Science Indicators, 1974*. Washington, D.C.

National Science Foundation. 1976. *National Patterns of R&D Resources, 1953–1976*. Washington, D.C.

—————. 1976. *Science Resource Studies Highlights*. Various issues. Washington, D.C.

—————. 1976. *Technological Innovation and Federal Government Policy*. Washington, D.C.

—————. 1974. *The Effects of International Technology Transfers on U.S. Economy*. Washington, D.C.

—————. 1972. *Research and Development and Economic Growth/Productivity*. Washington, D.C.

Organization for Economic Cooperation and Development. 1975. *Patterns of Resources Devoted to Research and Experimental Development in the OECD Area, 1963–1971*. Paris.

—————. 1975. *Changing Priorities for Government R&D*. Paris.

Pavitt, Keith. 1974. "'International' Technology and the U.S. Economy: Is There a Problem?" In *The Effects of International Technology Transfers on U.S. Economy*. Washington, D.C.: National Science Foundation, pp. 61–76.

Peterson, Willis L. 1967. "Returns to Poultry Research in the United States." *Journal of Farm Economics*. (August).

Rosenberg, Nathan. 1976. "Thinking About Technology Policy for the Coming Decade." Prepared for the Joint Economic Committee, U.S. Congress. Washington, D.C.

Solow, Robert. 1957. "Technical Change and the Aggregate Production Function." *Review of Economics and Statistics*. Vol. 39, No. 3 (August), pp. 312–320.

Stern, Robert M. 1975. "Testing Trade Theories." In *International Trade and Finance: Frontiers for Research*, Peter B. Kenen, ed. Cambridge, England: Cambridge University Press, pp. 3–49.

Stewart, Charles T., Jr. 1972. "A Summary of the State-of-the-Art on the Relationship Between R&D and Economic Growth/Productivity." In *Research and Development and Economic Growth/Productivity*. Washington, D.C.: National Science Foundation, pp. 11–19.

Terleckyj, Nestor E. 1974. *Effects of R&D on the Productivity Growth of Industries: An Exploratory Study*. Washington, D.C.: National Planning Association.

—————. 1960. "Sources of Productivity Advance." Ph.D. dissertation, Columbia University.

Tri-Association Ad Hoc Committee on IR&D and B&P. 1974. *Technical Papers on Independent Research and Development and Bid and Proposal Efforts*. Washington, D.C.

U.S. Department of Commerce. 1972. *Technology Enhancement Programs in Five Foreign Countries*. Washington, D.C.

U.S. International Economic Report of the President. Various years. Washington, D.C.

Vanderslice, Thomas A. 1976. "The Vital Need for Technology and Jobs." Speech presented to the Executives' Club of Chicago, November 5, 1976.

Vernon, Raymond. 1966. "International Investment and International Trade in the Product Cycle." *Quarterly Journal of Economics*. Vol. 80, No. 2 (May), pp. 190–207.

Weisbrod, Burton A. 1971. "Costs and Benefits of Medical Research: A Case Study of Poliomyelitis." *Journal of Political Economy*. Vol. 79, No. 3 (May/June), pp. 527–544.

National Planning Association

NPA is an independent, private, nonprofit, nonpolitical organization that carries on research and policy formulation in the public interest. NPA was founded during the great depression of the 1930s when conflicts among the major economic groups—business, farmers, labor—threatened to paralyze national decision making on the critical issues confronting American society. It was dedicated, in the words of its statement of purpose, to the task "of getting [these] diverse groups to work together . . . to narrow areas of controversy and broaden areas of agreement . . . [and] to provide on specific problems concrete programs for action planned in the best traditions of a functioning democracy." Such democratic planning, NPA believes, involves the development of effective governmental and private policies and programs not only by official agencies but also through the independent initiative and cooperation of the main private-sector groups concerned. And, to preserve and strengthen American political and economic democracy, the necessary government actions have to be consistent with, and stimulate the support of, a dynamic private sector.

NPA brings together influential and knowledgeable leaders from business, labor, agriculture, and the applied and academic professions to serve on policy committees. These committees identify emerging problems confronting the nation at home and abroad and seek to develop and agree upon policies and programs for coping with them. The research and writing for these committees are provided by NPA's professional staff and, as required, by outside experts.

In addition, NPA's professional staff undertakes research designed to provide data and ideas for policy makers and planners in government and the private sector. These activities include the preparation on a regular basis of economic and demographic projections for the national economy, regions, states, and metropolitan areas; the development of program planning and evaluation techniques; research on national goals and priorities; planning studies for welfare and dependency problems, employment and manpower needs, education, medical care, environmental protection, energy, and other economic and social problems confronting American society; and analyses and forecasts of changing national and international realities and their implications for U.S. policies. In developing its staff capabilities, NPA has increasingly emphasized two related qualifications—the interdisciplinary knowledge required to understand the complex nature of many real-life problems, and the ability to bridge the gap between the theoretical or highly technical research of the universities and other professional institutions and the practical needs of policy makers and planners in government and the private sector.

All NPA reports have been authorized for publication in accordance with procedures laid down by the Board of Trustees. Such action does not imply agreement by NPA Board or committee members with all that is contained therein unless such endorsement is specifically stated.

FRED KORTH
Korth and Korth

PETER F. KROGH
Dean, Edmund A. Walsh School of Foreign Service,
Georgetown University

*EDWARD LITTLEJOHN
Vice President–Public Affairs, Pfizer Inc.

WILLIAM J. McDONOUGH
Executive Vice President,
International Banking Department,
The First National Bank of Chicago

WILLIAM G. MITCHELL
President, Central Telephone & Utilities
Corporation

EDGAR R. MOLINA
Vice President, Ford Motor Company; Chairman
and Managing Director, Ford Latin America,
S.A. de C.V.

KENNETH D. NADEN
President, National Council of Farmer Cooperatives

*RODNEY W. NICHOLS
Vice President, The Rockefeller University

WILLIAM S. OGDEN
Executive Vice President, The Chase Manhattan
Bank, N.A.

JAMES G. PATTON
Menlo Park, California

*WILLIAM R. PEARCE
Corporate Vice President, Cargill Incorporated

CALVIN H. PLIMPTON, M.D.
President, Downstate Medical Center

S. FRANK RAFTERY
General President, International Brotherhood of
Painters and Allied Trades

RALPH RAIKES
Ashland, Nebraska

WILLIAM D. ROGERS
Arnold and Porter

STANLEY H. RUTTENBERG
President, Ruttenberg, Friedman, Kilgallon,
Gutchess & Associates, Inc.

*RICHARD J. SCHMEELK
Partner, Salomon Brothers

*LAUREN K. SOTH
West Des Moines, Iowa

*J.C. TURNER
General President, International Union of
Operating Engineers, AFL-CIO

THOMAS N. URBAN
Executive Vice President, Pioneer Hi-Bred
International

GLENN E. WATTS
President, Communications Workers of America,
AFL-CIO

*LLOYD B. WESCOTT
Hunterdon Hills Holsteins, Inc., Rosemont,
New Jersey

BLAND W. WORLEY
President, American Credit Corporation

*Executive Committee member.